# 我为绿地做软装

I DO SOFT DECORATION FOR GREENLAND GROUP

袁旺娥　康振强　编著

華中科技大學出版社
http://www.hustp.com
中国·武汉

# 序言 PROLOGUE

## 绿地的风姿

袁旺娥、康振强两位新锐设计师在该书中，以图文并茂的形式，向我们展现了他们以绿地集团作为软装艺术设计平台的最新软装创意成果，撩开了软装这个在中国姗姗而来的新兴行业的面纱。全书的内容收录了他们丰富实践与潜心思考的成果。从这些丰富的实战案例中，我们会惊喜地看到他们是如何将自身的美学观融入其软装设计的理念与造物的建构中，并由此带来独特的尚美品质与意境营造。这本专集的众多设计案例，就如同无数个晶莹闪光的七彩石，折射出两位新锐设计师丰富的生活感悟和富有想象力的现场创作能力，也最大程度地展现了他们的才思和美学生活理念。

十年前，袁旺娥、康振强分别毕业于中国美术学院视觉传达系和国画人物系。在美术学院的学术围墙里，他们以极大的热情对中外设计史及设计前沿的最新成果做了系统的研究和梳理。走出学校后，他们又以极大的热情投身到丰富的生活中，并聚焦软装设计领域，潜心研究，勇于实践。经过大量实战的设计与理性思考，他们在软装这个领域逐渐成熟，并建构起自己的美学理念下的软装概念和各种丰富的表现形态。他们能够基于中国的哲学思想和宇宙观来确立人的生存和家居美学，并以此为原点向当代生活的软装设计形式发散，传达出契合东方美学和独特个性意识的软装设计，使传统的文化和美学活化在今天，为我们带来了诗性、平和的新空间。

在这本《我为绿地做软装》图书即将付梓出版之际，我作为康、袁两人大学母校的学长，为他们在社会生活领域的专业实践所获取的真知和成就而感动，也更感到自豪！

愿他们未来的道路无比宽广。

安滨 博士

中国美术学院继续教育学院院长、教授

2014年10月15日

# Elegant Demeanor of Greenland Group

Here the new and vigorous designers Yuan Wange and Kang Zhenqiang in the book of "I Do Soft Decoration for Greenland Group" to show us their creative soft decorations by using Greenland Group as their design platform. With their latest creative achievements reveal the secrets of soft decoration which is the emerging industry in China. The contents of this book contain a wealth of practical and concentrated thinking achievements. From various practical cases we will be surprised to see how they are using aesthetics to penetrate their design concept and create a special beauty designs and atmosphere. The book collects numerous designs, which show us a variety of specific cases, just like countless shiny colorful stones which can reflect these two wonderful designers' rich life experience and creative designs, and also can utmost show their imaginative and aesthetic philosophy of life.

Ten years ago, Yuan Wange and Kang Zhenqiang graduated from the China Academy of Fine Arts in visual communication department and traditional Chinese painting personages department respectively. In this academic wall, they did many researches on Chinese and foreign design history and the latest designing achievements. After leaving school, with great passion they put their heart into the real life, and focus on soft decoration design field, with concentrating on studies and courage to practice. After a lot of practical design and cerebration, they grown up gradually in soft decoration design area, and build their own design concepts and a variety of performance under the guidance of aesthetic philosophy. And they can establish people's lives and home aesthetics base on the Chinese philosophy and world view, and regard this as original point to extend to the different forms of soft decoration under the contemporary life, and conveying the Oriental aesthetics and unique designs and making the traditional cultural and aesthetic come alive, bringing us the poetic, calm and joyful new space.

At the time of publishing the book of "I Do Soft Decoration for Greenland Group", as the college president of their graduated university, I feel very proud and happy for the achievements and truth they get in the field of professional practice of social life.

Dr. An Bin
President and professor of the college of continuing education of China Academy of Fine Arts
October 15th, 2014

# 前言 PREFACE

## 做软装的美学使者

转眼，从事软装行业已经8年有余。从懵懵懂懂地接触，到殚精竭虑地与各位行业同仁共同推进这一细分行业的发展，有太多的话要说，但真正提笔，又好像无从下笔。有一点可以肯定的是，2006年开始接触软装至今就一直认定这个领域，即使“软装”这个新颖的名词至今仍有争议。

确切来说，软装是从家装开始起步的。当初，是尊敬的余工余静赣先生把我们带上了软装这条路。涉足该行业之初，几乎没有多少可参考的资料，国内也几乎没有可借鉴的案例，但中国先富起来的精英界已经显现了对软装市场巨大的需求。这是个发展空间巨大的新领域，充满了瓦砾与荆棘，也充满艺术创新与市场需求。对软装知之甚少的我们，与当时国内第一批软装设计师一起，不断摸索、不断探求，摸着石头过河，积极开拓并积累有益经验。在前进途中，有些公司倒下了，有许多设计师离开了，而有些坚持者得到了。回想起来，我们能坚守到现在，追根溯源，离不开自己的兴趣爱好。对于我个人来说，由于从小就对美术情有独钟，对绘画、手工很感兴趣，经常折腾家具摆设，尤其对色彩很敏感，喜欢研究、分析色彩，这些都为后来自己在软装设计这个行业奠定了一定的基础。后来随着对软装的逐渐认识和个人经验的积累，在做项目时，我喜欢用色彩为客户打造适合他们的空间，以热情与奔放诠释生命的真谛，这往往都能得到客户的认可。这同时说明了在事业中，把生活和工作、兴趣和爱好结合起来才是最幸福的。而老康（康振强）跟我一样，原来一直学的是绘画，一直觉得做一名职业画家才是正事，但懵懵懂懂、跌跌撞撞摸索到今天，他也发现原来做软装设计这个服务行业是蛮有乐趣的。因为热爱，所以一切付出都是幸福的。我们一心扎在这个行业里刻苦钻研，倾心投入。也有朋友不理解，问：“软装行业所涉及的产品如此琐碎，工作上又非常烦琐及庞杂，没日没夜废寝忘食，是什么支撑你们坚持到现在？”我想说，还是热爱。因为热爱，所以付出的同时其实已经得到；因为热爱，所以我们经常可以超脱纯商业利益去追求艺术与生活最大的契合度；因为热爱，所以每一个出色的案例都能给我们带来成就感……这一切，都足以让我们一直坚持下去，并乐在其中。

值得一提的是，我学的是油画和视觉设计，老康学的是国画人物，即使专业不同，但艺术出身给我们在进行软装陈设时带来了很多创意和灵感。与其他人不同，我们作为同一个学校毕业的夫妻档，从学校到工作，从恋爱到结婚，无论在工作上还是生活上都相互促进、相互扶持，人生目标和未来规划趋于一致。从业至今做过很多地产项目，也获过多个奖项，其中在2011年的“筑巢奖”中获得的陈设艺术银奖，日本无印良品设计总监原研哉还亲自为其写了评语。即使这些年我们在软装行业取得了一些成就，但是我们一直没有忘记初衷，在经营公司的同时不断对软装学术进行研究，向“做软装的美学使者”这个目标靠拢，把喜欢的行业作为一个实现个人价值的平台，希望做一些贡献，以期为这个行业尽一份绵薄之力。

这些年来，我们做了许多案例，摸爬滚打地积累了一些经验，前两年也陆续参编了一些软装类的书籍，如《软装设计教程》《跟着大师学软装》《软装设计手册》等。我们从去年就开始筹划《我为绿地做软装》这本书了，之所以选择绿地集团的名字作书名，是因为在合作的众多地产商中，近几年与绿地集团合作得较多。有时候去艺术院校讲课，常有学生会问：“这个软装怎么做啊？”就一个简单的问题，却能让我们心潮澎湃。所以本书第二章“汉意堂软装操作流程详解”部分，具体阐述了软装设计的每一个步骤，其实就是我们把自己的经验拿出来与大家分享，无论是对同行还是即将从事这个行业的人，希望自己的浅见能给大家提供一些帮助。

在本书出版之际，特别感谢以下友人：

安滨博士，邓小鹏教授，陈妍教授，余静赣先生，马仁俊先生，Murad博士，于岩（美籍华人），John Pauline（美国），汉意堂全体同事。

2014年10月10日于广州

# Aesthetic Messenger of Soft Decoration Design

Time flies, it has been more than eight years since I get into the field of soft decoration industry. Actually, from zero to really familiar with this industry, I have too much to say, but when I pick up a pen, I find, I don't know how to start really. While one thing for sure is that since 2006 the first contact with soft decoration I have chosen this concept, even if the term of soft decoration is still new and controversial now.
To be precise, soft decoration begins with home improvement. At first, I was guided to the soft decoration road by respected Mr. Yu Jinggan. At that time there was no reference for us, and almost no cases we can learn from, but the rich elite circles in China have appeared a huge market demand for soft decoration. This new filed, not only filled with rubble and thorns, but also full of artistic innovation, market demand and a variety of possibilities. At the beginning we knew little about soft decoration, but with the first batch of domestic soft decoration designers together, we continued to explore and accumulated useful experience. On the way forward, some companies fell, some designers left, and some people get success. Now when I remember this, I think the reason why I still insist on this field it must be my hobbies. For me, I have been interested in the fine arts, painting and handmade since I was young. I often furnished and decorated home myself. I am very sensitive for colors, and like to research and analysis colors. All those hobbies laid a certain foundation for me to engage in the soft decoration industry in the future. Later, with the gradual accumulation of knowledge and experience, I like to use colors to create a comfortable space for customers with passionate and enthusiastic interpretation of the essence of life, and my designs are always accepted by customers. From this I know that it will be the happiest thing if we can combine our life and work with interests and hobbies. Like me, Mr Kang had been trained as a painter, and certainly think to be a professional painter was the right way for us, but have been struggling and exploring till now, he also found that the soft decoration service industry was also full of fun. Because of love, all hard work will become happiness. We put our heart and soul into this industry and enjoy ourselves. Some friends do not understand and ask: "the soft decoration involve so many trivial products and much complex work, without day and night work so hard and even forget food and sleep, what support you to insist on till now?" I want to say: "it's still love! Because of love, we have actually gained a lot while trying; because of love, we can combine purely commercial interests with art and life; because of love, every good case we designed can give us a sense of accomplishment ... so all this will be enough to make us persist and enjoy it.

Actually, it is worth mentioning that I studied painting and visual design, and Mr. Kang studied traditional Chinese painting personages, even if the majors are different, the artistic background brings us a lot of creativity and inspiration when we begin to design a soft decoration project. Unlike others, as a husband and wife team studied at the same university, from school to work, from fall in love to marriage, whether work or living, we promote each other and support each other, and with the same life goals and future plans. After engaging in this field, we did many real estate projects, and we also won a lot of awards, like "Nest Award" in 2011, we got silver award for the best art display. The design director of Japanese MUJI, Kenyahara also wrote a comment for us personally. Even though we have got some achievements all these years, we haven't forgotten the original purpose, while running the company we continue to do academic researches on soft decoration, to get closer to the goal of being an "aesthetic messenger of soft decoration design". And make our favorite industry as a platform for the realization of personal value, and hope to make some contribution to this industry.

When it comes to the purpose of publishing this book, in recent years, we do lots of cases and accumulate some experience. Two years ago we started to participate and wrote some soft decoration books, such as "Soft Furnishing Design Guide" "Learn Soft Decoration Designs from Master" and "Soft Decoration Manual". We began planning this book from last year. And the reason why we choose Greenland Group as partner is that we have lots of cooperation with them in recent years. Other reason is that when we go to give lessons at some art school, we are often asked by student about how to design this soft decoration, such easy question always make us feel the thrills. So we give the answer in the second chapter "The Operating Processes of Soft Decoration", it specifically addresses every step of soft decoration designs. In fact, it is just the experience we want to share with people and we hope that our experience can help someone in the same field or some persons are going to engage in this industry.

At the time of publishing the book, I want to thank some friends here:
Dr. An Bin, Prof. Deng Xiaopeng, Prof. Chen Yan, Mr. Ma Renjun, Dr. Murad, Nick Yu (Chinese American), John Pauline (American),
All colleagues of Haniton.

Wanda
October 10th, 2014, Guangzhou

# 目录 CONTENTS

CHAPTER ONE

## 汉意堂与绿地的前世今生
## THE RELATIONSHIPS BETWEEN HANITON AND GREENLAND

CHAPTER TWO

## 汉意堂软装操作流程详解
## THE OPERATING PROCESSES OF SOFT DECORATION OF HANITON

# 3

## CHAPTER THREE

## 汉意堂案例展示
## HANITON CASES

CHAPTER ONE

# 汉意堂与绿地的前世今生

# THE RELATIONSHIPS BETWEEN HANITON AND GREENLAND

左：黄浩罡　中：袁旺娥　右：康振强

文/ 黄浩罡　Author / Huang Haogang

转眼间，我从一位设计人转向行业媒体已经多年，与汉意堂的康振强、袁旺娥夫妇也相识好几年了，可以说是看着汉意堂一步步发展起来的。近年来，汉意堂一直在为绿地地产提供软装系统的策略，出品精彩之作。我很好奇他们是如何与绿地集团合作的，今天来到汉意堂的后花园，就着一壶康振强刚从云南带来的普洱茶，话匣子就此打开……

How time flies! As a designer I have turned to the media industry for many years, and have been acquainted with Kang Zhenqiang and Yuan Wange couple for so long. In another words, I have been witnessing their development since then. In recent years they have been providing soft decorating strategies for Greenland Group, and producing so many wonderful designs. I am curious that how they cooperate with Greenland Group. With this question, I visit the Haniton's back garden. With a pot of Puer tea brought from Yunnan by Mr. Kang, the chat was beginning...

（以下，黄浩罡=黄，康振强=康，袁旺娥=袁）

( Below, Huang Haogang = Huang, Kang Zhenqiang = Kang, Yuan Wange = Yuan )

# 汉意堂眼中的绿地
# Greenland in the Eyes of Haniton

黄：合作源于相识和选择，我们一生中会遇到很多人，若能遇到真知，收获就会很大。我们没有办法选择出身，但是可以选择职业和合作伙伴，所以我想了解汉意堂是如何成为绿地集团的合作伙伴的？又是谁选择了谁呢？

康：我们接触绿地集团其实是一个偶然。我某日在浏览手机时，无意中看到绿地集团西安分公司的一个关于软装招标的信息，就和他们联系上了，之后甲方就让我们提供相关的文件和软装案例。令人意外的是，对方很认可我们的设计理念。其实我们就是这样通过公开的渠道获取信息，按照常规的手续公平竞标而建立合作的。

Huang: Cooperation comes from acquaintance and choice. We will encounter a lot of people in our life, and if we can meet the real knowledge at every stage of our growth the harvest will be enormous. We have no right to choose our parents, but we can choose a career and partners, so I want to know how Haniton becomes a partner of the Greenland Group. And who choose who?

Kang: Actually, cooperating with Greenland Group is an accidental chance. Oneday when I was browsing webpages by mobile phone, accidentally saw a soft decoration inviting public bidding information which is issued by Xi'an Branch of Greenland Group, so I contacted them immediately, and then they let us provide relevant documents and soft decoration cases, surprisingly, they approved our designs concept. In fact, we obtained information through open channel like this, and conducted fair competitive bidding according to regular procedure to set up cooperation.

黄：我想开始时你们应该是以一个普通设计参与者的身份介入的，随着时间的推移，双方在角色、关系和合作方式上发生了某些化学反应。你们的合作能够持续这么多年，双方的默契已经达到一个度了吧？当时你们是怎么了解绿地集团的需求的，并且义无反顾地选择他们作为长期的合作伙伴的呢？

康：是啊，很默契！可能刚开始的时候连绿地集团也不是很清楚自己的软装系统应该如何定位，我们是通过深入的调研分析以后才决定合作的。这也是我们的工作流程。对于准服务对象我们一直坚持前期要做充足的沟通和调研，而且我们的角色定位是合作伙伴而不是产品供应商！知己知彼才能百战百胜嘛。比如我们曾经专门研究过绿地集团的品牌标志，绿地集团logo左边是房子，右边是树，而绿地的英文“green”盖住了树，也就是说绿地集团要保护生态并把自然引入人的生活之中，说明绿地集团充分理解并注重生态的可持续关系，正如他们的广告词“绿地，让生活更美好”，这也是该企业文化最核心的东西。绿地集团从品牌形象到理念都透露出它是一个有社会责任感、有长远抱负的企业。这样的合作伙伴当然是我们非常认同的。也正因为有了这样的理念，绿地集团近年来也走向了国际市场。2013年，绿地集团在洛杉矶筹备开发一个大项目，同时在澳洲、欧洲、韩国等地也都有战略布局。这反映出绿地集团不仅是中国的，更是世界的。

就合作而言，汉意堂很注重甲方的实力，但比这更重要的是甲方对项目软装设计的理解和运作方式。在与绿地集团合作的过程中，我们发现绿地集团设置了协调统筹市调、建筑、硬装、软装等环节的协作部门，因此在项目立项期，这些专业的参与者都有机会一起参与研讨，共同商议最佳的开发方案，这在与其他的地产商合作中是比较少见的。汉意堂的品牌基因是艺术，我们希望把每个项目都当成艺术品去做，但我们希望合作伙伴要能理解并支持，只有双方相互尊重相互欣赏，才会出品精彩。

这些年，我们也接触了其他的地产企业，他们对软装的重视度不够，有些是模块化操作。而绿地集团对软装要求独具个性，强调每个项目的地域文化价值和独特的设计价值。

Huang: I think at the beginning you should sever as a general design participant in this cooperation, but with the time going by, the roles, relations and cooperation ways of the two sides have been changed. So your cooperation can be sustained for so many years, the two sides must have reached a degree of understanding, right? Then how did you understand the needs of Greenland and without hesitation to select them as a long term partner?

Kang: yes, we have a tacit understanding! Maybe even Greenland Group themselves were not very clear how to position their soft decorating system at the beginning. Before deciding cooperation, we had done some deeply research and analysis. This is our work processe, for prospective clients we have been insisting to do plenty of communication and research in advance, and our role is a partner rather than a product supplier! As the saying goes that he who has a thorough knowledge of the enemy and himself is bound to win in all battles. For example: We have specifically studied the logo of Greenland, you see, on the left there is a house, on the right there is a tree, and the English word of " green " covers the tree, that means the Greenland aims to protect the natural ecology and introduce it to human life, and that also means they are fully understanding of ecological ethics and focus on sustainable relationship. As their advertisement says: " Greenland, create better life " , this is the core content of their corporate culture. From the brand image and to the concept, they all show that the Greenland is an interprise with social responsibility, and it is a long-term vision enterprise. Of course, this will be a partnership we are highly desirable. It is because of this concept, Greenland recently gets chance to enter into the international market. Last year Greenland planned to develop a large project in Los Angeles, as well as Australia, Europe, South Korea and other countries also overall strategie arrangement. This reflects that the Greenland is not only owned by China, it also belongs to the whole world.

In terms of cooperation, Haniton Decoration Design pays more attention to the customer's strength, but more important than that is the understanding and functioning of the soft decoration designs on the project. In cooperation with Greenland, we find that they set up some extension departments to connect with the market research, construction, hard decoration, soft decoration and other aspects, so in the period of establishment of the project, we can have the opportunity to participate in the discussion and obtain the best development program, this is relatively rare in the cooperation with other real estate companies. Haniton is based on art, we hope to do every project as a work of art, but this needs your partner's understanding and supporting, like love, one - sided love will not have result, only respecting and appreciating each other can get wonderful result.

Over the years, we have contact with other real estate companies, their attention to the soft decoration is not enough, and most of them are using modular operation. While the requirements of Greenland are personalized, they emphasize on the value of regional culture and unique design value.

黄：看来康先生很有感触啊！正如你所说的，评估和研究合作对象应该是建立合作的第一个阶段，例如需要评估和研究产品（楼盘）的建设质量和服务，甚至包括它的社会形象。任何一个企业都应该具有社会责任感，企业有没有回馈社会和民众也是评判其实力的考虑因素。第二阶段就是我们要用什么方法和工具，做什么样的准备来获取合作意向。这些动作可以看出一家设计公司是否具有策略性，比如进行数据分析的意识和方式、评判对方价值观的方法、获取项目合作意向及其他附加值的能力、团队内部的排兵布阵等。两位能否具体谈谈对绿地集团的了解？

康：那就先说说我们对绿地集团产品线的了解吧。我们对绿地集团产品线的定位表现在了解、选择和服务上。绿地集团涉及了房地产、能源、金融等产业，其中在房地产领域，有以下几个业态。

第一个是高层或超高层建筑。比如即将建成的武汉绿地中心，它是亚洲第一高楼。在这个项目中，绿地集团很好地运用了国外的先进环保生态技术和理念。绿地集团在超高层建筑开发领域积累了丰富的经验，开发规模和水平都属于全国同行的领先地位。以企业梦想刷新城市高度，目前建成和在建的超高层建筑已经有23幢了，其中4幢的高度位列世界排名前10！绿地集团的项目遍布全球，成为所在城市首屈一指的新地标。

第二个是住宅。国内做住宅的开发商很多，但是绿地集团在住宅上所呈现的产品线的层次很丰富，除了像“海珀”这一类型的高端系列以外，还有“云都会”“香树花城”等其他有特点的系列。

第三个是集商业、办公、酒店等功能于一体的现代服务业城市综合体。绿地集团整合了国内外一流的商业资源，与很多商业巨头形成战略合作关系。他们正在逐步搭建综合高、中、低端产品系列的立体式商业地产开发运营平台。同时，绿地集团也在参与全国各地城镇化发展的过程中凭借开发实力和产业运营能力，为各地量身打造了结合“生态、产业、居住、休闲”等功能于一体的24小时活力新城。绿地集团的“产城一体化”的模式可以说是开创了产业发展与城市发展良性互动的新格局，这也成为绿地集团进军世界200强企业的发展引擎和全新增长点。

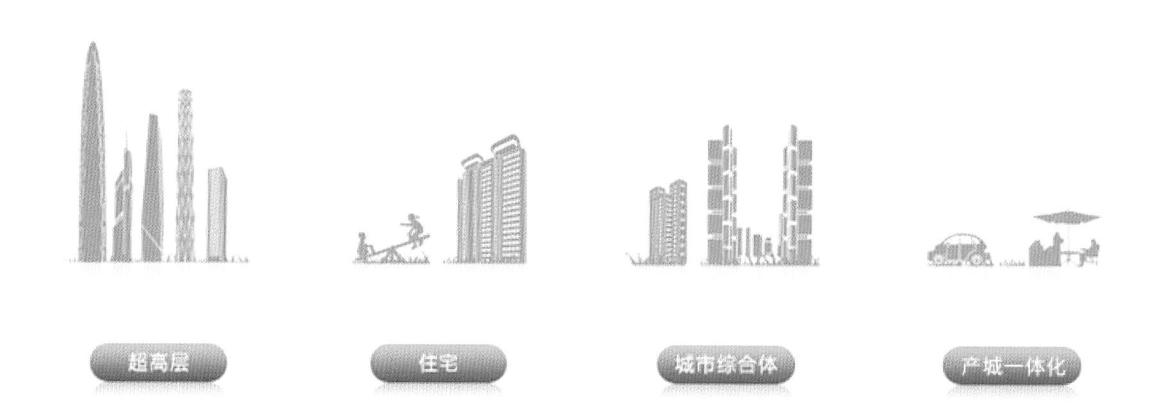

Huang: It seems Mr. Kang has very mixed feelings now! As you said, evaluating and researching cooperators is the first step to establish partnership, for the product (real estate) in the construction of quality and service, even including its social image, every enterprise should have a social responsibility, whether to contribute to the community and the public or not, also becomes a reference of enterprise evaluation. The second step, that is, which methods and tools can be used, and what kind of preparation can obtain cooperation intention. These actions can be seen whether the company has a strategic, such as the awareness and the way of data analysis, the value judgment method, the ability to obtain cooperation intention and other value-added and other formations within the team. Could you specifically talk about the understanding to the Greenland?

Kang: Ho-ho, then let us talk about what we know about the Greenland product line. Our position on the Greenland product line which is shown by understanding, selection and service. The business of Greenland covers real estate, energy, finance and other industries, and in the real estate field, there are several formats:

The first is high-rise or ultra-high-rises building. For example, the forthcoming completion of the Greenland Center in Wuhan, close to completing will be the highest building in Asia, and they make good use of the foreign advanced eco-technologies and concepts. Well-experienced in building ultra-high-rises and leading in project size and quality nationwide, Greenland Group constantly refreshes the city skyline with its corporate vision and becomes the leading operator in developing ultra-high-rises in China. There are currently 23 ultra-high-rise buildings completed or under construction by the Greenland, four of which are among the top ten highest buildings in the world. All of them have become the new landmarks in different cities around the world.

The second area is residential. There are many residential developers in China, but the Creenland Group' product line has rich levels, in addition to "Hysun" this type of high-end series, also including "Cloudy Metropolis" "Cree and Flower of City" and other features series.

The third area is the modern service industry complexes which integrate commerce, office, hotel and other functions. Moreover, the Greenland Group actively integrates the first-class commercial resources at home and aboard, and forms strategic partnership with many business giants to gradually set up a stereoscopic service platform for commercial real estate operation with high, medium and low-end product series. Backed up by its development strength and industry operation capability, the Greenland Group has built the tailor-made 24-hour dynamic new cities that integrate "ecology, industry, inhabitancy and leisure" and other functions in its participation in urbanization around China. Greenland's model of "City with Industrial Integration" has created a new pattern where industrial and urban development interacts with each other.

袁：绿地集团有如此多的产品线，但我们现在主要接触的是绿地集团的住宅类项目产品。做乙方，一定要理解甲方的产品线划分所代表的意义。汉意堂在新项目定位时，会综合分析项目的各个方面，比如楼盘区域、售卖人群定位、售价、项目整体优势、园林、建筑、室内设计、风格取向、甲方预算，以及每个部分的权重是多少，等等。将这些信息回馈到软装设计时，我们对于软装的定位就会非常清晰，包括材质的运用、色彩的运用、空间尺度感、品质感的呈现等都会根据前面的分析来判断。

Yuan: The Greenland Group has so many product lines, but we are mainly dealing with some residential projects. To be party B, one should be sure to comprehend the meaning that the division of party A's product lines represent. While conducting new project orientation, Haniton will analyze each aspect of project by synthesis, such as property region, customer orientation, price, overall advantage of projects, landscape, construction, interior design, style orientation, party A budget, etc. The percentage of each section clears up the orientation of soft decoration when returning feedbacks to soft decoration design, including the application of material and color, the presentation of spatial scale and quality, which will all be decided according to the foregoing analysis.

黄：与绿地集团合作的过程中，绿地集团给你们留下的最深刻的感受和认识是什么？

袁：我觉得无论是做设计还是布场，整个过程都很有挑战性。

康：我觉得最深刻的感受是绿地集团对软装专业的尊重。我们与很多开发商都有合作，但在甲乙双方的思维角度方面，大多数甲方还是居高临下，完全是按照采购-供应商的模式竞标操作。虽说甲方可能会了解一点软装知识，但毕竟不是很精通，我觉得尊重了软装专业，只会对甲方更好，作品也会更精彩，如果把最终呈现效果放在次要位置，而纠结于一些不重要的琐碎事物的话，很容易喧宾夺主，事倍功半。

Huang: What are the most impressive feeling and cognition that the Greenland Group has left to you?

Yuan: I think whether providing design plans or decorating sites, the whole process is full of challenge.

Kang: I think the most impressive feeling is the Greenland respect our profession. We have worked with many developers, but as the different perspectives of both parties, most of the first parties still show the condescending attitude and full compliance with buyer - supplier mode through competitive bidding way. Although some people in the first party know a little about soft decoration, after all, they are not good at it. I think show some respect to the professionalism, can only be better for the first party, and the works can be more exciting. If we pay more attention to the unimportant trivial things and put the final effects in the second place, it is easy to turn things upside down. People look likes busy, but the important things are not yet done.

# 康、袁二人的软装心路历程
# Soft Decoration Experience of Mr. Kang and Ms. Yuan

黄：作为一个设计行业的媒体人，我一直有个疑问，行业内有的将软装叫作“陈设”，在专业命名上不统一，社会的辨识度也就不高，这应该与软装的发展历史是有关的，对此我想请教一下两位。

康：我认为正确的称谓应该是“软装”，用“陈设”这个词来描述显得过于单薄了。我们是从2006年开始做软装设计的，当时整个行业都处于萌芽阶段。开发商最初是不需要做样板房来卖楼的，但随着行业竞争激烈，一些开发商就找一些花店、家具店来做简单的装饰，市场也很认同。花店和家具店自然做不到专业系统的服务，很多环节很难把控。所以开发商开始向一些从事硬装的设计公司发出这方面的合作邀约。不久，这种需求和要求越来越多，慢慢地就形成了我们这个行业工种。现在软装设计已经延伸到酒店、会所及其他商业空间了。

Huang: As a media editor in design industry, I always have a question, someone called this soft decoration, others call it display. If there is no unified title in spreading, the social recognizable will be not good. I think it should have some connections with the history and development of soft decoration, for this, I would like to ask advice of you two.

Kang: I think the proper title should be soft decoration, using the word of display to describe it seems too thin and simple. We started to do soft decoration design from 2006, when the entire industry was in its infancy. At that time developers did not need to do model room to sell their houses, but with the intense competition in the industry, some developers began to find some florists and furniture stores to do simple decoration, also this is accepted by the market. However florists and furniture stores couldn't provide professional and systematic services and it was difficult to control in many areas. So some developers began to ask interior design companies to do this job, and the needs and requirements became more and more, then slowly formed our industry. Now soft decoration has been extended to hotels, clubs and other commercial spaces.

黄：你们觉得目前人们对软装的认知是怎样的？是否如硬装一样在追求风格？

康：我认为风格的形成应该是一种地域性文化和审美的体现，比如中式风格、地中海风格、东南亚风格，就是这一区域人们的一种生活方式的积累。每一个地域因为地理位置、宗教信仰、人文文化等因素在漫长的岁月积累中形成了独具特色的风格。从应用设计的层面来说，风格更多是一种族群或个人精神层面深层次的需求。如果你喜欢浪漫，可以选法式风格；如果你喜欢朴素简约，可以选北欧风格或日式风格。风格本身不重要，重要的是您的选择。很多人觉得大房子必须要呈现欧式，显得气派，其实不然。人们在选择的时候要回归到内心和现实的需求，自己喜欢并且适用才是最重要的。

袁：我去美国、法国和意大利等地游学时，感觉当地人对自己的审美非常自信，但国内很多人自己喜欢什么都不确定，总是喜欢询问周围的朋友。其实每个人都有自己的沉淀，适合别人的不一定适合你，认识自我非常重要。当然，在装修房子时，考虑的是整个家庭，从一个人到一个家，怎样平衡各个家庭成员的喜好点非常重要。从这一点上也说明一个问题，即对设计师的要求也就更高了，所以一定要为服务对象的生活选择一种合适的“表述基点”。

Huang: What kind of perception towards soft decoration do people have at present? Is that the same in pursuing styles as hard mounted?

Kang: I think the styles come from regional cultural and aesthetic expression, such as Chinese style, Mediterranean style, the Southeast Asian style, which is the people's accumulation of lifestyle in that region. Every locations, religions, cultures can form a unique style after long years of accumulation. Everyone's taste is different. Here can only show the deep-seated spiritual needs of an ethnic group or individual. Like if you love romantic, you can choose French style; If you like plain simple, you can choose Nordic style or Japanese style. In fact, style itself is not important thing, while the most important thing is your selection.

Yuan: When I studied in United States, France and Italy, I found that the people in there were very confident with their aesthetic appreciation and taste, but our people always like to ask advices of their friends around. Actually, everyone has their own characteristic, and a style that suits others may not be suitable for you, it let us know self-understanding is so important. Of course, in the decoration of the house, the whole family will be taken into consideration, from one person to a family, how to balance the preferences of each family member is very important. From this point we can see, the requirements of the designers also will be higher, so designers have to select an appropriate basic point for the customer's life.

图为汉意堂设计总监袁旺娥在美国、意大利游学

黄：你说到“选择一种表述基点”，让我想到一位设计大咖，杭州内建筑设计公司的沈雷。他曾提出了一个概念——平民时尚。因为他认为中国无贵族，他重视的是“窃窃的喜、偷偷的乐”，寻求的是平民的这种静谧的小愉悦。所以在他的很多作品里都以“平民时尚”为设计基点，作品中无不体现了很有趣的情景故事性的空间，有一种心灵和时光交互的动态感，这也是他表述的意识形态。软装也一样，它更注重人在现实生活中的心理因素，表述的基点确实很重要。

康：是啊，软装对于家庭太重要了，一方面要功能合理，另一方面要顾及视觉美感及秩序的营造。一个人的美感提升是持续积累的，但现在很多人分不清美与丑，搞不清雅与俗。清华美院的潘吾华教授曾经提出一个观点，她说中国人以前是“文盲”，现在是“美盲”，我非常认同这个观点。往深层次追究原因，问题应该是出在基础教育和社会环境上。在荷兰阿姆斯特丹的国家美术馆，我亲眼见证了四五岁的小孩被老师带着看世界名作，他们不是走过路过，而是在一幅画面前一坐就是几十分钟，老师从美学、历史、地理、宗教等很多角度和孩子们互动交流，真是让人羡慕！

袁：我们现在所能做的就是从自我做起，生意和职业是一个方面，更多的是想做一个合格的美学使者，引导更多人认识美及美的应用。

图为汉意堂总经理康振强在北欧考察

Huang: When we talk about " basic point " , I remember a design master in Interior Design named Shen Lei, who proposed a concept - Civilian Fashion. He believes that there is no nobility in China, he puts emphasis upon the secret joy and happiness and wants to show this quiet pleasure of civilians. In his works you can find the point of " Civilian Fashion " everywhere, and also all his works can create some very interesting stories of space. There is a sense of the dynamic interaction of psychological and time, that is the concept he wants to express. Soft decoration is also focus on capturing people's real feeling. It is the reflection of the importance of the basic point.

Kang: Yes, soft decoration is so important to a family. On one hand, we have to provide a logical function; on the other hand, it needs us to create a visual aesthetic feeling and order. To improve one's taste is a long-term work, but now a lot of people can not distinguish beauty and ugliness, and confuse with elegant and vulgar. Professor Pan Wuhua in Tsinghua Academy of Arts & Design had been made a point that Chinese people used to be " illiterate " but now we seem to be "beauty blind " and at this point I couldn't agree with him more. To study it further, you will find the problem comes from the basis of educational and social environment. In the National Gallery in Amsterdam, I witnessed some four-year-old children were looking at the world masterpieces with accompany of their teachers, they did not just pass by, but sat in front of the paintings tens of minutes, teachers and children would interact in many aspects such as aesthetics, history, geography, religion. That scene was really enviable!

Yuan: What we can do now is to start from ourselves. Besides business and career, we should be qualified aesthetic messengers, and guide more people to appreciate beauty. We should be teachers just like the way we call each other in the company, we call each teacher.

黄：我想，从软装的美学意义还可以延伸出一个概念—— 新软装模式下的新型人居关系。在过去，三姑六婆经常互相串门，分享生活的经验，而现在大家都是关上门来过自己的生活。能不能通过个性化的软装，让邻里亲朋愿意串门，分享各自的生活呢？这是不是一个值得软装设计师深思又很有趣的话题呢？

康：是啊，软装是能够改造家庭关系和优化家庭环境的。现在城市的人居关系存在着很大的问题，我们作为软装设计师也在考虑这一点。软装呈现出来的东西，如果让人很愉悦，那就使得人与人之间愿意相处。首先家里很舒适、很漂亮，你回到家里心情很好。因为你家沙发很舒服，你家的花花草草很漂亮，别人也会喜欢到你家做客，愿意与你沟通。其实就是要让这个家的主人从精神层面上得到缓解和释放，比如能拥有阅读、听音乐、谈心或洗个香薰浴这种生活方式，那你就愿意回家了。我觉得家就是爱的道场—— 夫妻之爱，长幼之爱—— 软装产品就是道具，先把道具设计好，大家就很容易入戏，就像喜欢看书就要有个像样的书架，喜欢听音乐就在床头柜放一个品质好的IPOD一样。

袁：我们去年在洛杉矶设立分公司，所以在那待了很长一段时间，没事的时候就去看房子。在这个过程中，我接触了很多美国的普通业主。看了他们的家后，我很惊讶。这都是一些很普通的美国人，也不是从事艺术或设计相关行业的，但居然从硬装到软装都可以亲手去完成，并且整体效果非常棒。往深层次挖掘，就回归到人家的教育上，在国内同样也看过好多别墅，从外面看似乎“高大上”，但进去后大都只能用“惨不忍睹”来形容。可能是因为我本来就是艺术专业出身，对有美感的事物比较敏感吧！我们希望去帮助更多的人把居住空间弄得更漂亮、更舒适，所以我们成立了一个意堂家居事业部，把汉意堂这些年积累的好作品整理出版，帮助社会民众去解决美学问题，做一个能落地的美学传播者。

康：嗯，关于袁老师所说的“能落地”，我举个例子吧！中国的传统文化中有非常灿烂的东方居住文化，特别是比较鼎盛的明清家具，在世界家具史上是浓墨重彩的一笔。但文化有精华也有糟粕，中国长久的封建制度在很多家具上都有体现，比如官帽椅，除了形式的借鉴外，所蕴藏的深层次内涵就是坐在这把椅子上要端着，四平八稳。坐在这种椅子上短时间内是没问题的，但时间长了，会非常辛苦。相对来说，坐在沙发里肯定会更放松。现在很多人喜欢买红木家具，家里有那么几件东西可以回味历史挺好的，但有些人的家里整屋都是红木家具。问他为什么，他告诉你30年后会升值到多少多少，但连他们自己都不会坐的东西花那么多钱买来又有什么用呢？家具的本质功能是从人体工程学的角度让人做任何事情时达到最佳的舒适度，我在七月游学北欧时发现，北欧的一些设计师在这方面做得还是很棒的。所以，我们真的要想想每天要用的东西是30年后能升值重要，看起来“高大上”重要，还是当下用着舒服最重要？

Huang: Interesting! From the aesthetic significance of soft decoration I think I can also extend it to a concept: a new habitat relation under the new mode of soft decoration. Take a look at our previous habitat relationship, we often visiting each other, sharing experiences of life. Now we are living our own lives behind the closed doors. So whether can we back to previous lifestyle, letting neighborhood visit relatives and friends and willing to share their lives, is it an interesting topic for soft decoration designers to consider?

Kang: Yes, it is. Soft decoration can play a role to improve and refine family relations. Now there is a very big problem with Habitat relationship of the citizens, which should be considered by our soft decoration designers. If soft decoration shows a pleasant state, it makes people willing to share. First the comfortable and nice home will give you a good mood when you get back home, meanwhile others will be willing to visit, because you have comfortable sofa and beautiful flowers and plants. In fact, soft decoration is just to make you feel and relaxed, enjoying a lifestyle such as reading a book, listening to music, talking freely and taking a bath , then you are willing to go back home. I think that home is dojo of love, with love of couples, love of elders and younger, soft decoration products as props which are designed in advance to make everyone easy to act into play, just as to have a presentable bookshelf if you like reading, or to put a high quality IPOD on the night table if you like listening to music.

Yuan: Last year, we set up a new branch office in Los Angeles, so I stayed there for a long time. When I was not busy, I went to see the houses. In this process, I met a lot of ordinary American owners and I was surprised by their home. Those were some very ordinary Americans. And they were not engaged in art or design-related industries, but actually they did the all decorations by themselves and there homes looked great. I beleved they were well-educated. I have seen a lot of villas in China, too, from the outside they "high-end, nice and great " , but after entering into the villas we can only use " tragic " to describe. Probably because of the art professional background, I am very sensitive to beautiful things. How to help more people to make their living space more beautiful and more comfortable is a duty for me. So we set up Haniton, collecting and publishing the good works which are accumulated for years, to help people to solve aesthetic problems and becoming real aesthetic communicators.

Kang: Now we talk about the phrases " can be used " which is mentioned before by teacher Yuan. Here is an example. In Chinese traditional culture, we cultivate a splendid oriental living culture, particularly the acclaimed Ming and Qing furniture, which leaving an Indelible mark in the history of world furniture. Culture, both essence and dross are contained in it. Chinese long feudal system can be reflected in many furniture, such as official hat chair, sitting on it is not comfortable, relatively, sitting on the sofa will certainly be more relaxed. Now a lot of people like to buy redwood furniture. When you ask them why, they will tell you how much the furniture will be 30 years latter, but I wonder if they will not use them why spend so much money to buy them. The essential function of furniture is to be convenient to use. I found some Scandinavian designers do this great when I studied in Northern Europe in July. So, we really have to think what is the most important thing for us, rising in value after 30 years, looking " high-end, nice, great " or comfortable at this moment.

黄：没错，物尽其用才是明智的，毕竟家不是博物馆。我觉得软装的功能还可以延伸开来，比如是否能解决人与人之间的关系问题？

康：是有可能的，这也是我们接下来要研究的一些课题，以后可以延伸开来研究。比如避免“夫妻七年之痒”的软装系统、解决隔代沟通方式的软装系统。我认为这就是软装行业的未来走向，从小处着手，引领老百姓改变一些不良的生活方式，引入更多积极的生活方式。比如，现在中国的孩子“衣来伸手，饭来张口”，我们希望通过软装提供一个空间能够让小孩爱学习并主动去收拾房间。我们要去做这些事情。另外，汉意堂未来还会结合智能系统让软装系统智能化，比如智能化的床，早上醒来可以看到一组睡眠的数据，包括睡眠质量、睡眠深度、睡姿等的分析，这都是令人兴奋的话题，我们很向往。

Huang: That's right. It is wise to make the best use of everything; after all, home is not museum. I think that the function of soft decoration can still be extended, with the example of whether it can solve problems of interpersonal relationship.

Kang: That is possible. This is also projects we are going to research next. These projects can be researched extensively such as soft decoration system of seven-year itch between couples, soft decoration system to solve inter-generation communication. In my opinion, this is the future trend of soft decoration industry, which lead people changing some undesirable life-style, at the same time bring in more positive life-style in some details. For example, today's Chinese children are spoild to take care of themselves in daily life. How does our soft decoration provide a space to attract children to study, to clean up their rooms actively? That's the thing we need to do. In the future, Haniton will use intelligence system in soft decoration, such as an intelligent bed which shows you a set of sleep data including sleep quality, analysis of depth of sleep, sleep position analysis, etc. These are exciting topics we are yearning for.

黄：软装设计师给予的是一种心理的满足、审美方面的培训和应用方面的落地。一个人在什么样的环境下长大，对他的素质和未来发展都有影响。搭配不同的饰品和家具起到的作用完全不一样。除了美育，软装设计师也一定要考虑到安全和健康问题，比如家具、饰品或是枕头的填充物会不会对孩子有副作用。那么，两位觉得一个软装设计师应该具有哪些素质或技能呢？

袁：前段时间，广东省陈设艺术协会组织学生在汉意堂上课时，就涉及这个问题。有一个组的学生做的是小孩房，在他们的作品里面，他们要考虑很多选材方面的问题，比如要考虑小孩的呼吸，用什么质地不容易起毛，不容易对小孩的呼吸道造成影响。我们不是单纯地考虑视觉，视觉只是软装方面应该考虑的一项，当然美观是必要的，但美观的同时要考虑房间中的边边角角，这也是我们对软装设计师的要求。软装设计师设计的作品必须要符合人性，让客户感受到自己的生活品质在提高，那么，客户在用这个产品的时候就会觉得，软装设计师给他带来了思考的空间。这也是软装从业者要具备或者思考的方面。

艺术家可以按照自己的想法把内心的东西呈现出来，无需顾虑太多。但商业设计师是一个服务者，我们在进行一个完整家庭的软装设计的时候，一定要了解男主人是谁，女主人是谁，什么职业，他们的生活方式是什么，生活中的角色是什么，也包括他们在社会中的身份。作为软装设计师，怎么让业主真正放松下来，让家变成真正的港湾，我们要对所有的家庭成员进行一个深度的调查，这也是软装从业者要具备或者要思考的，当时绿地集团也是看重我们这一点。

Huang: Soft decoration designer can provide a psychological satisfaction, training appreciation of aesthetic and realization of application value. A person grows up in what kind of environment has an impact on the quality and future of him. With different decorations and furniture play a completely different role, in addition to aesthetics, soft decoration designer must also take into account of the safety and health issues, such as the padding of furniture, decoration or pillow will have side effects on children. So, what kinds of qualities or skills soft decoration designer should have?

Yuan: Some time ago Guangdong Association of Art Crafts and Decoration Industry organized some classes in Haniton, and we discussed this topic in the class. A group of students chose to design a kid room, in which they did a lot of work on the selection of materials, for example, what texture has little impact on children's respiratory tract. We do not simply consider the vision effect, of course, beauty is necessary, but we should also take every corner into account, this is the requirement for our soft decoration design. Products designed by soft decoration designers must be consistent with human nature, and making the owners feel their life has been improved, so when owners use this product, he will feel, designer gives his thoughts. It is soft decoration designers must have and think.

Artists can follow their own ideas to express something inside, without worrying too much, while the designer is also a commercial service provider. When we conduct a whole soft decorating for one family, be sure to understand the owners, what kind of job they do, what kind of lifestyle they have, what is their roles of life, including the identity of the society. As soft decoration designer, how to make the owners feel relax and let the home become a real harbor, firstly we have to know all family members and make a depth investigation, it is important for soft decoration designers and this is the reason why we are selected by Greenland.

康：现在的软装设计师鱼龙混杂，真正理解软装所承载的丰富内涵的很少。究其原因，一是这个行业毕竟是一个新兴行业，大家都在摸着石头过河，没有前车之鉴；二是从业者接受的教育不全面，并且很多人对这个行业的认知还有很大的分歧。比如从做室内设计过渡到软装设计的，他们会认为一个软装设计师最重要的是要理解硬装；很多人认为，国外设计师不会分得那么细，在国内也没有必要独立出来叫软装设计师； 还有很多原来经营各种软装产品的，也转行做软装设计。没有基础的审美教育和系统的设计从业经验，如何能做好设计呢？

其实软装设计师的门槛非常高。一个软装设计师的成长，最主要的是积累，从最基本的美学基础到建筑、室内装饰设计史、家具史、人体工程学、材料及流行趋势，以及心理学等。虽然很多的知识会和室内设计师重叠，但在真正的工作过程中，室内设计师面对的工作主要侧重于空间规划，硬装的各种材质运用。如果让硬装设计师也完全负责软装的工作，工作量之大、知识面之广，他是非常难做到的。软装设计师正好完成了硬装设计师的后续工作，让空间得以完整呈现。

当你从事了软装设计，慢慢就会变得难以自拔。一朵小花、一个雕饰、一块布料，背后那些数不清的故事，让你不断地探索。我和袁老师已经研究八年了，但我们知道，别说八年，就是八十年也无法完全参透软装的无穷奥秘。它已经变成我们生活的一部分及生命的一部分。漫步在欧洲、美国、日本的街头，总觉得肩上的使命很沉重。希望中国人的脚步慢一点，多去体会生命成长路上的点点滴滴，让美好时刻伴随左右。在这个过程中，我们软装人能奉献一点点微薄的力量，已经非常开心了。

就像汉意堂的口号“享受软装，享受艺术”所说的那样，做一个幸福的软装人！

Kang: Now the levels of soft decoration designers vary greatly, only few of them really understand rich connotation carried by soft decoration. The reason is, first, the industry is an emerging industry, everyone is trying, and there is no example. Second, practitioners accept incomplete education, and there is a great different awareness in the people of industry. For example, some interior designers who change to do soft decoration, they would think a soft decoration designer should pay much attention to the hard-mounted. And many people it is no specific division in foreign design field, so there is no need to separate out soft decoration designer. Meanwhile many soft decoration products sellers also switched field to become a designer. If we do not have basic aesthetic education and experience, how can we create a wonderful design?

In fact, the barrier to entry the soft decoration field is quite high. The growth of designers, depending on accumulation, including the most basic aesthetic foundation to architecture, interior design history, furniture history, ergonomics, materials, trends, as well as psychology and so on. Although a lot of knowledge will overlap with the interior designer, in the real process the work of interior designers focus primarily on spatial planning, a variety of hard-mounted materials used and if hard-mounted designer is also fully responsible for soft decorating work, it will be very difficult to do because of a large workload and extensive knowledge covering. The role of soft decoration designer is to make up for the continuation of hard-mounted work, and let the space can be fully present.

When you are engaged in soft decorating design field, you will find it hard to pull out. A flower, a carving, a piece of cloth and a countless stories behind them, make you constantly explore. Teacher Yuan and I have studied this for eight years, but we know that, even study for another more 80 years we still can not fully understand the endless charm of soft decoration. It has become a part of our lives, and a part of life. I always felt the heavy burden on my shoulders when I was walking on the streets of Europe, USA and Japan. Wish Chinese people can slow down our pace, to record dribs and drabs during our life and growth, and let the happy time stay with us. If we soft decoration designers can make a little contribution to this process, it will our honor.

Just like the slogan of Haniton says: " Enjoy soft decoration, enjoy art ". Be a happy soft decoration designer!

CHAPTER TWO

# 汉意堂软装操作流程详解

# THE OPERATING PROCESSES OF SOFT DECORATION OF HANITON

- 方案设计 Program Design
- 采购、制作 Procurement and Production
- 陈设展示 Furnishings and Presentation

常常留恋于和谐、舒适的空间，或奢华、或简约、或极简主义，尔后惊叹设计师独到的原创的视角和那双灵巧的手。我们总是贪婪地感受着这一切，时常好奇它是怎样产生的……带着各种疑惑和不解，我们走进了汉意堂的后花园，一窥软装从构思到实现的全过程……

I often love to wander in a harmonious and comfortable space, some luxury, some simple, and some minimal, and then I will be surprised at the designer's unique perspective and skillful hands. We always enjoy all of these greedily and wonder how it is produced... With a variety of questions and doubts, we walked into the Haniton's back garden, to find out the entire process of soft decoration from concept to realization...

随处可见的灵感之源——凋花、旧墙、陶器、雨、新绿，白石。

Finding inspiration everywhere - sent flowers, from the old wall, pottery, rain, green, white stone.

# 方案设计
# Program Design

# 一、获取甲方资料

# I. Getting Information of Party A

**1. 软装设计要求**

如果甲方是地产商，则甲方已经根据整个楼盘的区域定位、目标客户、销售卖点等对项目本身有了自己的规划和定位，比如指定哪种软装风格，如何有效地弥补户型或硬装的不足等。如果项目是酒店，则甲方更多的是从酒店的整体定位来要求软装设计。如果项目是办公室或商业空间，则软装设计需要从实用角度出发。

**2. 硬装设计效果图**

有了甲方已经确认的效果图后，软装设计师对硬装的设计手法、方向定位、优缺点等都会有很直观的认知。

**3. 硬装平面图及施工图**

通过平面图，设计师可以清晰地了解实际工地中的各方面信息。施工图是非常必要的，因为只有通过施工图，设计师才能清楚地知道每个空间的施工细节，特别是立面方面，墙壁是何种处理手法，窗的高度和层高等。另外，施工图提供了详细的尺寸，方便设计师在做软装方案设计的时候正确把握空间中大件物品的尺度，如家具。在有些户型中可能几厘米的误差就导致某些家具放不下，带来很多后期的麻烦。

1. Soft Decoration Design Requirements

Generally, when party A finds a cooperating soft decoration company, they have to know more about what they want before. Like a real estate developer, they will make there own planning and positioning in advance according to the regional location of the real estate, target customers and selling points, then they can decide what kind of style they want and how to make up for the lack of space or hard-mounted effectively. If the project is a hotel, soft decoration design should focus on the hotel's market positioning. If the project is an office or a commercial space, the soft decoration should be funtional.

2. Hard-mounted Design Renderings

After confirmed of the renderings by party A, the processes and final effect will be just like the rendering shown. So when we have renderings, we can know directly such as the hard-mounted design techniques, orientation, advantages and disadvantages, ect.

3. Plans and Construction Drawings Hard-mounted

Through hard-mounted plans, we can clearly know the every aspect of the actual construction site. And the construction drawing is also very necessary, because only by this we can clearly see the construction details of each space, especially the elevation drawing, it will show us the processing method to designing the walls , window height, floors and other construction plans can only be reflected by elevation plans. Moreover, construction plans has provided us with detailed size, to help us in controlling item sizes when we designed the soft decoration for the space, such as the furniture sizes. Sometimes only a centimeter of error would cause some furniture does not fit, and it will bring a lot of trouble.

## 二、对项目进行详细分析
## II. Analysis of the Project in Detail

**1. 项目行业特色**
① 地产类（售楼处、样板房）
② 酒店类
③ 商业空间（餐厅、KTV、美容院、办公室等）

**2. 项目硬装情况**

**3. 对业主的分析**

1. Characteristics of the Project Industry
① Real Estate (Sales Offices, Model Room)
② Hotel
③ Commercial Space (Restaurant, KTV, Beauty Salons, Office)

2. Situation of Hard-mounted

3. Analysis of Owners

通过以上的分析，我们就可以列出一个详细的软装项目设计任务书了，格式如下：

By the analysis above, we can list a detail task book of soft decoration, shown as following:

**（1）项目概况**

项目名称：

项目地点：

项目面积：

甲方设计负责人：　　　　联系电话：

我公司设计负责人：　　　联系电话：

**（2）设计要求**

A. 设计内容和范围（家具、灯饰、窗帘、饰品、床品、地毯、挂画）

B. 设计定位

a. 情景主题：

b. 整体项目主题：

c. 具体空间主题：

d. 风格定位：中式、东南亚、现代、欧式、新古典、其他

C. 设计进度计划

**（3）设计成果**

A. 设计进度计划书

a. 提供概念设计成果时间

b. 提供方案设计成果时间

c. 提供材料板时间（家具布料及木饰面板）

d. 提供家具白胚完成时间

B. 初步设计概念图册

a. 人物背景、爱好设定，如男主人的职业、爱好等

b. 平面优化布置图

c. 家具、灯具等方案彩图

d. 深化设计图

e. 家具清单

f. 灯具清单

g. 窗帘清单

h. 饰品清单

i. 床品清单

j. 地毯清单

k. 挂画清单

l. 家具布料及木饰面样板

**（4）其他要求**

（1）Project Overview

Project Name:

Location:

Project Area:

Design Manager of Party A:　　　　Tel:

Responsible person in our company:　　　　Tel:

（2） Design Requirement

A. Design content and scale (furniture, lighting, curtain, ornament, bedding, carpet, and painting)

B. Design Positioning

Subject Theme:

Project Themes:

Space Themes:

Style Definition: Chinese style, Southeast Asian style, Modern style, European style, Neoclassical style, others

C. Design Schedule

（3）Design Results

A. Design Schedule

a. Providing conceptual design results

b. Providing program design results

c. Providing material plate (furniture fabrics and Wood decorative panels)

d. Providing furniture blank completion time

B. Preliminary Design Concept Album

a. Background and hobbies setting, such as the career of owner, hobbies and so on

b. Optimization of layout

c. Scheme color pictures of furniture, lighting and so forth

d. Deepen design drawings

e. Furniture list

f. Lighting list

g. Curtain list

h. Decoration list

i. Bedding list

j. Carpet list

k. Paintings list

l. Furniture fabrics and Wood decorative panel model

（4）Other Requirements

创意追随着灵性，虚幻与现实交织错落，阳光、青春、过往、未来、音乐……

Creative following with inspiration, illusion intertwined with reality, sunshine,youth, past, future and music ...

## 三、方案设计阶段
## III. Program Design

设计来源于生活，但又高于生活。针对项目的不同，我们要选择不同的设计主题。设计主题是贯穿整个软装工程的灵魂，是设计师表达给客户“设计什么”的概念。

Design comes from life, but beyond life. For different projects, we will choose a different design theme, which is the soul through out the entire soft decoration project, and is also an expression of the concept of design.

### 1. 地产商售楼处及样板房住宅项目

这一类型的项目要从体验式生活方式的设计主题去做。原因是：住宅项目面对的都是老百姓、普通业主，当他们去买房时，他们对未来充满了憧憬，并不断地规划自己有了新房后的新的生活方式，这个时候，我们恰好展现了一幅幅美丽的生活场景。

### 2. 酒店

酒店的软装设计一般比较强调文化性。酒店的文化又是多方面的，如酒店历史传承、酒店地域特色文化、酒店整体星级档次定位等，在设计时设计师就不能沉迷于某一个领域去思考，而是多方面、宏观地去思考。酒店的选材要非常讲究，在考虑材质、大小、色彩、空间的同时赋予准确的文化定位，这才是一个优秀软装方案设计。酒店的很多软装部分都是定制型的，无论是雕塑还是摆件、挂画、窗帘，都要非常考究，经得起世界各地旅客的考量，属于画龙点睛之笔。没有文化方面的设计考量，就会使空间显得空洞、空乏、肤浅、草率等。

1. Sales offices and model room

The theme of the project will be fit the requirement of experience. Because residential projects are facing the ordinary people and home owners, they are always longing for the future and planning of new lifestyle when they go to buy a house. At this time, we just show some examples of depicting beautiful scenes of life for them.

2. Hotel

The soft decoration of hotel will pay more attention to culture, while this culture here contains many aspects, like historical heritage ofction of materials should be very particular about, texture, size, color, space and then give accurate cultural positioning, This is an excellent soft decorations reflect. Most of the soft decorations are costumed in hotel, like sculptures, ornaments, paintings, curtains; they must be very elegant and become a key point of the space, and can bear test of travelers allaround the worlds. Without culture some improper aspects will be shown, such as without content, lacking in soul, superficial, and hasty, etc.

## 四、如何组成一套完整软装设计方案
## IV. How to Finish Whole Set of a Soft Decoration Design

**软装设计方案文本组成**

◆ **文化篇** ◆

设计目标

项目分析

设计思想

客户群定位

风格定位

人物主题设定

色彩分析

材质分析

◆ **方案篇** ◆

平面图

空间明细设计图

**Soft Decoration Design Text Composition**

Culture Article

Design target

Project analysis

Design ideas

Customer positioning

Style positioning

Character theme Setting

Color analysis

Material analysis

Design Proposal

Plans

Design Drawings of Space Details

**1. 设计目标**

一直以来，汉意堂都在执著于国际设计语言的本土化叙述，专注于空间价值的充分开掘，旨在于科学控制成本的基础上为投资者创造最大化经营利好。对于所服务的项目，我们的设计目标是：

为本区打造国际风范——文化精品板房；

向内外埠展示设计精品——绿地国际品牌。

**2. 项目分析**

进行项目客户群分析主要要考虑当地地理环境和文化定位，要分析当地现有资源，考虑项目设计与周边环境相结合，最后得出该项目的规划定位。

1. Design Target

All the time, Haniton is persistent in introducing wonderful international designs to china, and focusing on the value of the space, aimed at creating the maximum benefit for investors on the basis of scientific costs control. Our design goals are:

To create international style and boutique culture space for you;
To show wonderful designs and the Greenland Group brand to the world.

2. Project Analysis

Project customer analysis mainly depends on the local geography and cultural positioning, to analyze existing local resources, considering the combination of the surrounding environment, finally reaching the conclusion of project planning and positioning.

**3. 设计思想**

设计思想来源于对设计目标、业态趋势、客户定位、风格定位、项目本体等的分析。（以某项目为例）

3. Design Ideas

It comes from the analysis of design goal, industry trend, customer positioning, style positioning, and project analysis and others. ( Taking one project as an example )

| 设计目标<br>Design Goal | 业态趋势<br>Industry Trend | 客户定位<br>Customer Positioning | 风格定位<br>Style Positioning | 项目本体<br>Project Analysis |
| --- | --- | --- | --- | --- |
| 文化精品房<br>Boutique Culture House | 综合体验场景感<br>Overall Experience Scene | 改善型、投资型<br>Improvement Type, Investment Type | 东南亚风格<br>Southeast Asian Style | SWOT分析<br>SWOT Analysis |

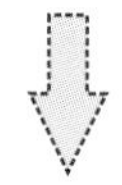

情境主题
Subject Theme

热带风情
Tropical Style

空间气质
Space Character

沉稳+异国+混搭
Peaceful + Exotic + Mix And Match

**4. 客户群定位、风格定位、人物主题设定**

项目的客户群是什么样的人？文化产业人士、商人、企业家、海归人士、政界商英、国企高管、私企高管？客户群定位需要考虑业主职业、身份、年龄、喜好、品位甚至经济实力等。

4. Customer Positioning, Style Positioning, Character Theme Setting

What kind of customers are they? Members of the cultural industries, businessmen, enterpriser, returned overseas people, politicians, the state-owned enterprise managers, or enterprises senior managers? We should consider the owners' professional identity, age, hobbies, tastes and even capacity to pay, etc.

风格定位往往是根据客户群的定位来设置的。软装项目的客户群通常已经有了一个良好的经济基础。客户或是希望拥有一个舒适大气、低调精致的生活环境，或是希望拥有一种奢华高贵的居家空间。因此，设计师可根据实际情况来决定，是做一种非常纯粹的软装风格，还是混搭型的软装风格。

Style Positioning: Style positioning is usually set up through the positions of customers. Designers consider this kind of customers as a crowd with good economic base; on the other hand customers expect a comfortable, low key and delicate living environment or a luxury and noble home space, which are all determined by the orientation of customers. Therefore, designers can create a kind of extremely pure soft decoration style or a kind of mixed soft decoration style according to practical situations.

**5. 色彩分析**

一个优秀的软装作品，最能够打动人的一般是色彩，所以一套作品中色彩具有无可比拟的重要性。同样的摆设手法，只是因为色彩变了，气质就完全变了。软装色彩遵循所有设计的色彩原理，一个空间要有一个主色调，一两个辅助色调，再搭配几个对比色或临近色，整个空间的效果就出来了。软装设计当中，设计主题定位好了之后,就要思考空间的主色系，运用色彩带给人不同的心理感受，灵活进行配色。

5. Color Analysis

Color is so important that when people are touche by an excellent soft decoration design most because of it. Even the same decoration will look different when color has been changed. Follow the principles of colors of all designs, in one space, there will be a main color, one or two secondary colors, and then with a few contrasting colors or similar colors, the effect of the entire space will come out. After deciding a design theme, we should think of the space main color, in the next, then according to the different feelings bring by colors, we can use them freely .

**6. 材质分析**

一个优秀的软装设计师一定是非常了解软装所涉及的各种材质的，不管是家具木料，还是油漆、布料、家纺、窗帘、陶瓷、玻璃等都要非常熟悉。不但要熟悉每一种材质的优劣，还要了解怎样通过不同材质的组合来搭配空间的风格。比如要打造一个清爽的蓝白经典地中海的风格，家具就尽量选择开放漆，木料尽量选择橡木或胡桃木，而布料尽量选择棉麻制品，灯饰的材质选择铁艺，这样搭配的空间就会把舒适、休闲、清新的地中海风格表现得淋漓尽致。

每一种材质都有其独有的气质，就像香水再香也不能多瓶香水混合用一样，一定要通盘思考整个空间，硬装的材质也要考虑在内。下面列举常用的不同材质的气质。

6. Material Analysis

A good designer must be very familiar with the various materials involved in soft decoration designs, including wood, paints, fabrics, home textiles, curtains, ceramic, glass, etc. Not only to know the pros and cons of each material, but also to understand by combining different materials to match the appropriate style for space. For example, to create a classic Mediterranean style in cool blue and white, the furniture should choose open paint, and oak or walnut wood; fabrics should use cotton cloth products; the material of lighting should select wrought iron, and such collocation will express the comfortable, leisure, and fresh Mediterranean style completely.

Each material has its own unique feature, just like even if perfume is so fragrant, we can't mix up different perfumes in one bottle. We must take the entire space into consideration, including hard-mounted materials. Here are some qualities of different materials.

| **材质** Material | **气质** Quality |
|---|---|
| 棉 Cotton | 有亲和力、舒适度非常高 Affinity and Comfortable |
| 麻 Linen | 自然、透气性强 Nature and Breathable |
| 丝绸 Silk | 细腻、柔滑 Delicate and Soft |
| 玻璃 Glass | 通透、清澈 Transparent and Pure |
| 水晶 Crystal | 晶莹剔透、品质感 Clear and High Quality |
| 不锈钢 Stainless Steel | 时尚、冷、酷 Fashion, Calm and Cool |
| 黑檀 Black Wood | 纹理美、高档感 Beautiful Texture and High-end Qulity |
| 陶器 Pottery | 自然、质朴 Nature and Naivety |
| 瓷器 Porcelain | 华贵、雅致 Luxury, Elegant |
| 铜 Copper | 富贵 Splendid |

物质相互的作用“爆发”惊人的能量，不同质感的融合展现出无穷的魅力，一盘好棋，两人对弈，四品香茗……

Different materials send out amazing energy when they combined, different textures show infinite charm when they mixed. Just like two people in a chess game, with a pot of tea, enjoying good moment...

为了使甲方能直观地感受到设计中所用的家具材质与色调，设计师往往会以展板的形式来予以说明。这不仅体现了设计师的专业度，而且展板本身也具有很强的美感，会给客户留下很好的印象，有利于方案的进一步实施。

In order to directly show the materials and colors used in the design to party A, designers often use such panels to explain, this can not only better reflect the professional degrees of designer, the poster itself also has a strong aesthetic, which will give customers a good impression and it’ s conducive to the further implementation of the program.

**7. 平面布局图**

一般来说，一个空间在建筑设计初期就已对空间的使用进行了合理规划，硬装设计部分也会对空间的平面做进一步的详细设计。到软装设计这一步，空间布局设计的余地一般就不是很大了，但也有一些大型的空间，如售楼处、宴会厅等，布局可以用多种形式，软装设计师就可以发挥了。只是在平面规划中，对家具尺度的把握要特别注意，比如一些客厅或主卧里面是放洽谈椅还是休闲沙发，整个体量感是完全不一样的，一定要根据实际空间来掌控。

7. Floor Plan

Generally, in the early architectural design, the rational planning of the use of space had been taken into consideration. The hard mounted will also think about the further detailed design of the floor plan. To soft decoration, there is almost no space to play in the layout part, but for some large spaces, such as the sales offices, restaurants, the layout can be various. Meanwhile, in the floor plan, we should pay attention to the scale of the furniture, such as placing Chat chair or leisure sofa in living room or bedroom can create a completely different feelings, so we must control the soft decorating design based on the real space.

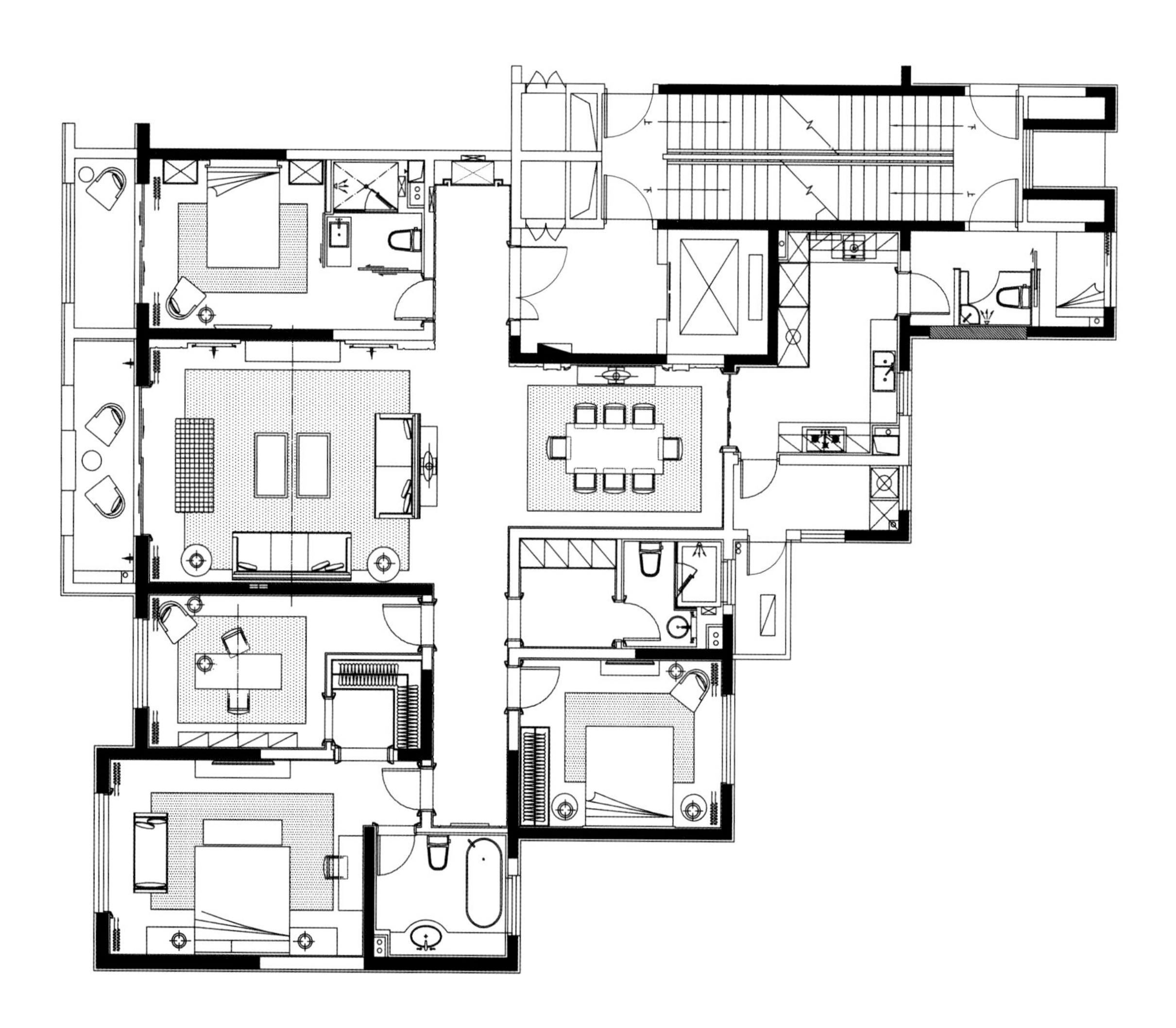

**8. 空间明细设计图**

软装的设计要更注重实用性。虽然设计师在做方案的时候都会融合一些情境图片来烘托氛围，但是软装设计毕竟最终要落实到实物上，所以在选择图片的时候，效果好是一方面，更主要的是选择的东西在以后的实施中到底能否采购或制作，既然是空间的明细设计，家具、窗帘、灯饰等所有软装物品在方案的设计中要有的放矢，重要的部分要展现出来，一些次要部分在报价清单里展现即可。下面以某一项目的空间明细为例。

8. Design Drawings of Space Details

Design of soft decoration should pay more attention to practicability, although we will fuse some contextual images to foil the atmosphere when processing scheme. However, soft decoration design eventually implements on real objects in the final, which leads that it is more important for chosen stuffs to be able to purchase or produce in future implementation than have a good effect when selecting pictures. Since it is detail designs for space, each type of soft decorative items in the design of project including furniture, curtain or decorative lighting should have a definite object in view with important parts revealed and some minor parts showed in quotations list. Take the following space details of one project as an example.

负一层吧台区

一层书房

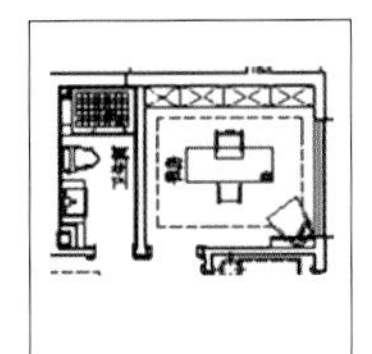

## 采购、制作
## Procurement and Production

合同签订后，接下来的事情就是要备货了。软装物品所涉及的种类庞大而烦琐，做好方案只是万里长征的第一步，有点停留在理念或纸面的意思。一个好的案例的产生光靠一个好的方案是远远不够的，如果没有配合好采购和摆场，最后的效果就会大打折扣。

### 采购前注意事项

如何采购，这在软装工作中也是非常有讲究的，比如采购顺序。正确的顺序是先预定家具，然后再预定灯具和布艺，最后才是其他。如果顺序不是按照这个方式，很容易会出现尴尬的局面。之所以先确定家具是因为家具在所有软装物品中通常是工期最长的，在甲方规定的时间内，如果前面的时间都耗费在采购一些饰品、画品等方面，那么后面就没有足够的时间进行家具制作，而仓促赶出来的家具可能会出现一些质量问题。

After signing the contract, the next thing is to prepare for the producting. Soft decoration items including many types of goods, good plan is only the first step to move, only a paper work stuck in the idea or concept. Without proper procurement and site furnishing, a case can not be successful just because of a good plan.

### Purchase Precautions

How to purchase is also very particular to the soft decoration companies, such as purchase orderly. Generally we first purchase furniture, and then lamps and cloth, finally are the others, otherwise embarrassing situation maybe appear in this project. The reason why we determine furniture first is that in the process of soft decoration, the production of furniture will take the longest period. In limited time, if the previous time spent in the procurement of decorations, paintings, etc., there will be not enough time for furniture production.

整个软装项目，成本最高的往往是家具部分，甲方对家具也是非常重视的，所以，采购家具在整个软装工程中也是非常关键的一步。

Furniture cost the most in the entire soft decoration project, so it is also very important to Party A. How to purchase furniture is a very critical step in the whole process.

## 一、家具

通常，家具的采购要根据客户需求来确定。一般来讲，工程类客户一般都会选择定制类家具，一些顶级商业客户会选择一些进口（比如意大利、法国、英国、美国）的品牌家具，而家居类的客户则比较中意国内各大家具卖场的品牌家具。

我们先分析一下几种类型的家具供应商。

### 1. 纯进口家具

进口家具往往积淀了数十年或上百年的文化，在国际上有很高的知名度，其材质和做工通常也会比较考究，当然价格也是非常昂贵的，动辄上万元至几十万。现在基本上由国内的代理商来操作，代表性的进口家具品牌有：FINDICASA、Girogetti、Baxter、Turri、C.G.Capelletti Ceccotti MDFItalia、Cassina、Zanotta Moooi、Magis Versace、RalphLauren、LIMA、BLUEMARINE……

## I. Furniture

Typically, the furniture procurement should be determined according to customers' demand. Generally, the project type customers usually choose custom furniture, some of the top commercial customers will choose some imported (such as Italy, France, UK, USA) furniture, while home customers prefer to some domestic brand furniture in the store.

First let us analyze some typical furniture suppliers:

1. Import furniture suppliers

Import furniture accumulates tens or hundreds of years cultural heritage, has a great reputation in the world, with a relatively good materials and workmanship, of course, the price is very expensive. Now they are often operated by domestic agents. Following are the representative of import furniture brands such as FINDICASA, Girogetti, Baxter, Turri, CGCapelletti Ceccotti MDFItalia, Cassina, Zanotta Moooi, Magis Versace, RalphLauren, LIMA, BLUEMARINE, etc.

采购特点：毕竟是漂洋过海过来的家具，所以货期一般都在两个月以上，有些畅销产品甚至可能要等半年。所以在采购此类家具时，一定要留有足够的采购时间，而一般地产公司或商业空间整个工程计划本来就非常紧张，因此往往很难用到纯进口家具。

Purchase Features: After all, the import furniture comes from oversea, so the delivery is generally in two months or more, and some of the best sellers have to wait for six months even more, when we need to purchase such furniture, we should make sure to leave enough time, and like real estate or commercial projects, time is too limited to purchase imported furniture only.

**2. 国内品牌家具**

随着时代的发展，近几年家具业的发展也是突飞猛进的，国内大型的家具卖场一直在增加，家具品牌也数不胜数。

采购特点：此类厂家大多数是流水线作业，批量生产供应全国的专卖店，本身定位是民用类家具比较多，90%以上的家庭客户会选择采购此类家具。其特点是厂家多，品牌多，选择余地大，但基本上都是按空间来规划一整套的产品，因此想打造非常个性的家具就比较难操作。

**3. 定制型家具厂**

为了满足工程类客户的个性化需求，现在特别是珠三角涌现了大量定制型家具厂，规模从几十人到几千人不等。

采购特点：非常适合工程类客户（售楼处、样板间、酒店、会所等），制作灵活，工期短，甚至一张椅子也可以定制。

定制家具流程：

（1）确定好尺寸，根据实际空间，一定要把尺寸核实清楚，有时候，家具就多几厘米便放不下，只能重做，造成重大损失。

（2）描述好细节，很多家具是有很多细节的，比如该用什么颜色的金箔，是玫瑰金还是香槟金，雕花的线条该多粗、多深，木料是用哪种木料，封闭漆还是开放漆，等等，很多细节都要详细告知家具厂。

| 项目：中泰潍坊项目B1-1户型别墅 | | 档案编号： |
|---|---|---|
| 内容：摆设台 | 尺寸：1900×480×550(mm) | 数量：1 |
| 位置：客厅 | 饰面： | 日期： |

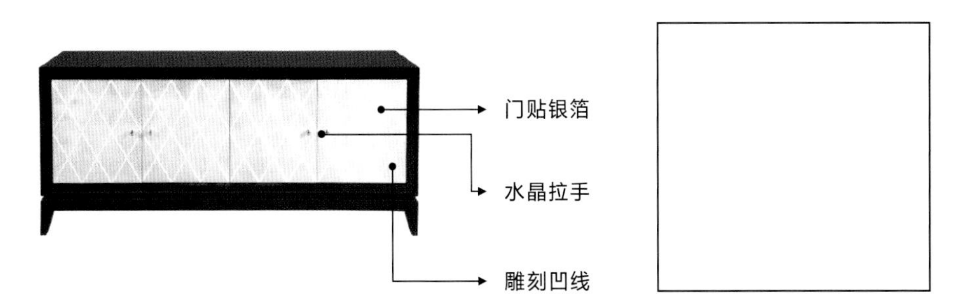


备注：以上图片仅供参考；除特别说明外，家具造型如示意图定做，所有家具以实木框架定做。

2. Domestic brand furniture in large furniture stores

Home furnishings industry has been developing rapidly in recent years and the domestic large furniture stores have been increasing. Now there are many furniture brands in China.

Purchase Features: Because most of these factories produce with assembly line work, mass productions supply stores nationwide, and the market positioning are more suitable for civilians, more than 90 percent of residential customers will choose to purchase this type of furniture. This kind of furniture is various in brands, but basically most of the products are classified according to the space sections, so it is difficult to create a personalized furniture.

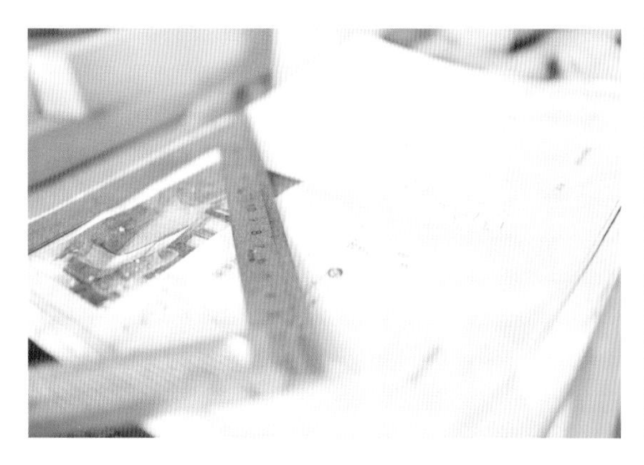

3. Customized furniture factory

In order to meet some large project clients' personalized needs, now emerged a large number of customized furniture factories in the region of Pearl River Delta, with a scale from dozens to thousands of works.

Purchase Features: this type of furniture is suitable for customers like (sales office, showroom, hotels, clubs, etc.) with flexible production, short period, and even a chair can be customized.

Custom Furniture Process:

(1) Sizes, according to the actual space, the sizes of the furniture must be determined and verified. Sometimes, even a few centimeters of error will make the furniture do not fit and cause a heavy losses.

(2) Details, there are many details of furniture, such as the color of the gold-leaf, rose-gold or champagne-gold, and the choice of carved lines, wood, paint and other details, the more the better. Please see the example below.

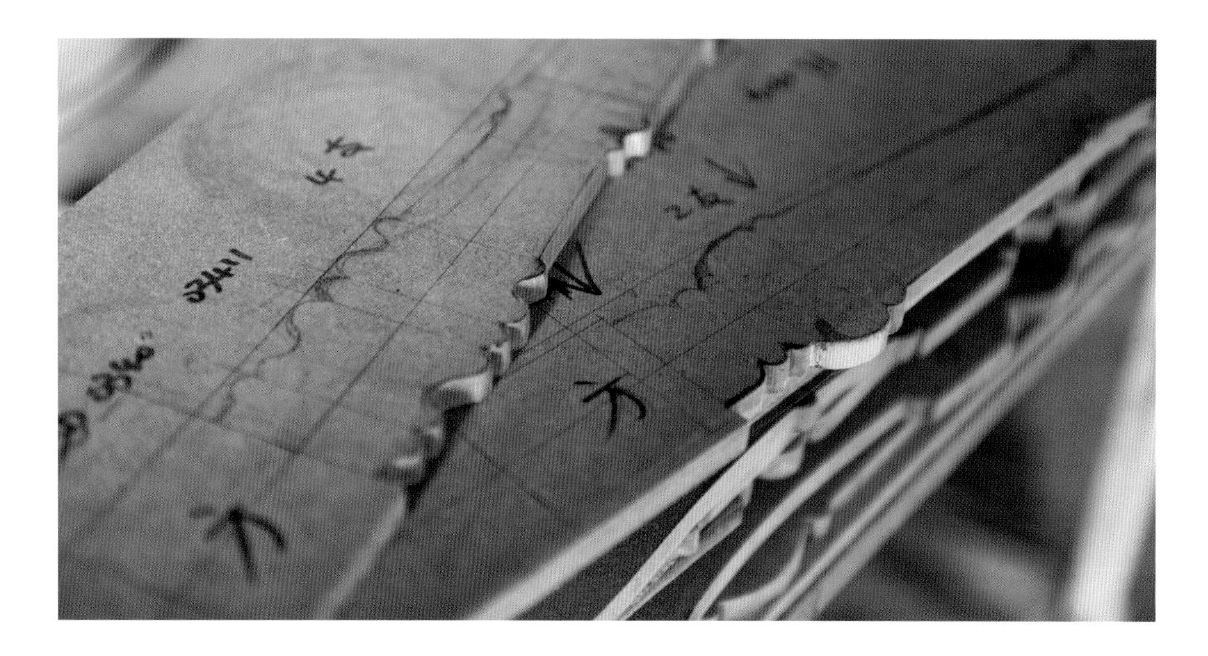

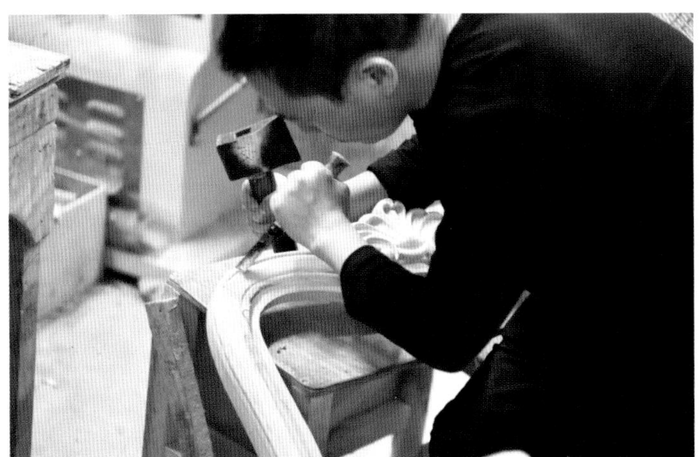

## 二、灯饰

灯饰的采购基本有两种方式：一种是按图样定制，一种是直接采购。

目前国内最大的灯饰生产基地为广东的中山古镇。在整个灯饰采购过程中，除了款式要与项目空间协调以外，还要注意材质。灯饰的成本计价基本都是按材料的价格再加上加工费算出来的。看上去效果差不多的材料，实际价格相差甚远。比如水晶就有进口（埃及）、国产A级、普通水晶等很多种，在采购过程中一定要和供应商确定用的是哪种水晶。一般来讲，纯铜的灯架价格非常昂贵，在灯饰中一般属于比较高档的产品了。

此外，灯饰采购时还要注意灯饰制作工期相对较长，因此在预定家具后就应该要预定灯饰了。

本页图片来源于网络 Photos from the Internet

## II. Lighting

There are two ways to purchase lightings, one is to custom lightings according to the patten, the other is to purchase directly.

Currently the largest lighting production factories are located in Zhongshan, Guangdong province. In the whole procurement process, in addition to coordinating style and space, be sure to clear the materials of the lightings. The cost of the lightings always base on the materials and processing technic. The product looks similar, but the prices are far more different, like Crystal, there are many kinds of types, import from (Egypt), the domestic A-class, ordinary crystal and other levels, so in the procurement process we must tell suppliers what kind of crystals we want. In general, the lightings using pure copper are very expensive, and it is relatively high-end product.

In addition to the precautions we are talking about the materials, we should pay attention to the relatively long production period. So after signing the contract with client, and purchasing furniture, we should order the lightings.

## 三、布艺

整个软装工程中，布艺部分非常烦琐，怎样在规定的时间内协调好窗帘、床品、抱枕等布艺部分也是非常重要的。

目前国内很多软装公司的布艺都是采购的，广州、深圳、杭州等都有这方面的市场。但是，为了把握整体的最佳效果，汉意堂公司所用的布艺都是公司的布艺中心自己设计及制作的。因为要做一个有特色、有主题、效果突出的空间，仅仅依靠采购是无法达到的，只有配备资深的布艺设计师，再配有完整的布料样板，才能真正发挥个性的设计。设计师可以把窗帘及床品、家具布料、抱枕等整个布艺部分，通过颜色、款式、材质、图案、制作手法等互相搭配、互相协调，做出设计师真正想表达的空间。当然，完全靠采购，并不是做不出好的作品，那就要考验设计师的整体把控能力了。

## III. Fabric

In the entire soft decoration project, fabric a is very complex part. How to complete the work of curtain, bedding, pillow and other fabric parts within required time is very important.

At present, a lot of soft decoration companies choose to purchase fabric products, and there are many makets like Guangzhou, Shenzhen, Hangzhou and so forth. However, in order to achieve the best results, Haniton has our own fabric design and production centers. Because if we want to create a distinctive, theme, and highlighting the effect of space, relying solely on purchases can not be achieved, only with a senior fabric designer and matched with a complete model, can really create a personalized design. For the parts of curtains, bedding, furniture fabric, pillow, etc., designer can use color, style, texture, pattern, production methods to coordinate with each other to make a space which designers really want to express. Of course, depending only on procurement, does not mean you can not create good work, but it needs the designer's ability to control the whole work.

下面介绍布艺采购的一些注意事项。

(1) 注意采购顺序。一般走进一个新空间当中，人们的视觉点往往分为四层关注度，分别为家具、窗帘、墙面、饰品与装饰类小件物品，所以要把家具布料部分放在第一位。确定了家具布料的颜色、材质、款式后，接下来就应该要考虑窗帘部分怎样呼应家具部分了。如果是卧室，床品部分应该和窗帘处于同等重要的位置，窗帘和床品确定后，再去思考抱枕、床旗、小布艺摆件等如何搭配整个空间。如果采购顺序不对，越到后面就会越难选择。

(2) 对布艺材质的了解和掌握。布艺有很多种材质，如常见的棉、麻、混纺、丝绸等，另外布艺材质也有进口面料或国产面料之分，在实际采购时，要根据具体要表达的空间来精心选择合适的布料，不同材质之间的搭配会呈现不同的效果。同样的花纹，采用不同的材质，其气质也会有很大的不同。

Here are some considerations for fabric procurement:

(1) Pay attention to the purchase order; generally when we walked into a new space in which people's visual point of attention is often divided into four layers, the first layer is furniture; the second is curtains; the third is the walls; the fourth is decorations and small items, so we should make the fabric part of the furniture in the first place, to determine the color of fabric, material, style, then we should consider how to make curtains and furniture match each other. In the bedroom, bedding should be as important as the curtains; after determining the curtains and bedding, we will think about the pillow, bed flag, small fabric ornaments, and how to decorated the entire space, If the purchase is in wrong order, it will be harder to make choices at last.

(2) Understand and master the materials of fabric; fabric composed by a variety of materials, such as cotton, linen, blends, silk, etc. And there are imported fabric materials and demonic fabric materials. In the purchasing, it needs us to carefully choose the right fabrics according to the specific spatial expression, and to create a very rich combinations by using different fabric materials. The same pattern, in different materials can bring a completely different feelings.

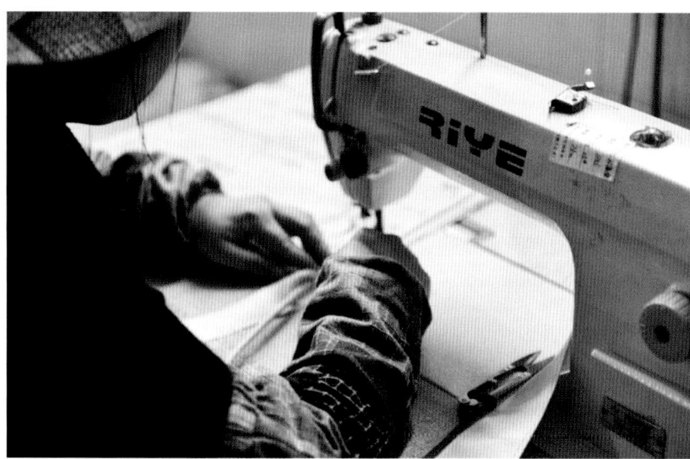
RIYE

RIYE

## 四、画艺

项目的不同决定了挂画种类的不同，某些高档酒店或会所会更倾向于采购名人字画。名人字画除了能极高地提升空间品位外，因其具有收藏价值而使得投资具有回报性。但是，这就要求业主在项目初期增加投入。

装饰性挂画已经不是传统意义上的挂画种类——不仅仅是国画、油画、水彩、水粉、漆画等之分——反而已经演化成了一种装饰艺术，所以更多的软装公司把挂画改成挂饰。其所采用的材料五花八门，比如皮革、不锈钢、珍珠、铜器、银饰、有机片、布艺等。具体来讲，材质不是主要的，重要的是靠这些材料营造怎样的整个空间的气质，包括其装饰手法及主题是否更好地协调整个空间。很多地产商的样板房项目比较喜欢装饰性挂画。但由于材料及手工等方面造价会稍高，因此，在采购的时候，一定要结合预算，如果选用此类挂饰，当初的报价就要适当高一些。

## IV. Painting

Different project determines different kinds of pantings we need, like some high end hotels and clubs they prefer to buy calligraphy and painting of famous persons, which will help to increase the taste of the space. Even more those works with their collectible value can bring a return on investment, relatively, this requires customers to invest a great amount of money at the beginning.

Decorative paintings, are not classified according to the traditional sense, not just Chinese painting, oil painting, watercolor, gouache, lacquer painting, etc. Those paintings have evolved into a kind of decorative arts, some of the soft decoration companies even use hanging ornaments instead of paintings, such as leather, stainless steel, pearl, bronze, silver, organic sheets, cloth, and other wide variety of materials. Specifically, the material is not the main important thing, the more important thing is to use the materials to create a certain space. Many showroom projects would be more like those decorations, but as the high cost of materials and other aspects of the manual, the price will be slightly higher. We should be sure to consider the budget, if we want to choose these ornaments.

对于画艺，我们通常会自己动手DIY，甚至装裱环节也会，这样更有利于实现自己的设计理念。在面对客户指定名家或者需求量大、报价低的时候，我们可以去市面采购或定制。

We usually make paintings, which are more suitable for our designs by ourselves. If the client wants some works of a specified master or has large demand, with lower price, we can purchase directly or choose to custom the products.

原创不在于用多昂贵的材料，而在于用心的设计。

Originality dose not depend on how expensive materials you used, but on how willing you want to create.

## 五、花艺

花艺师会根据软装设计师需要的风格搭配不同类型的花，样板房和会所多用于参观和销售，为保障作品最终呈现的效果，设计公司多采用仿真花和干枝。

## V. Floral

In accordance with the styles which soft decoration designer need, the florist will arrange different types of flowers. For example, like model rooms and clubs are the space for visits and sales, in order to achieve the best effect, designers will use some artificial flowers and Mikie in the design.

## 六、饰品

现在的家居饰品行业厂商已经多如牛毛，生产的各类材料和款式的家居饰品数不胜数，挑选的余地非常大，但经过调查会发现，大多数的家居饰品流入了工程类项目，如样板间、会所、酒店、家具卖场。

## VI. Decorations

Now there are too many home decoration manufacturers, with various kinds of types and materials, the choices are endless, but through the investigation we will find that the majority of home decorations are used in some large projects, such as showroom, clubs, hotels, and furniture stores.

最大的批发市场应该是深圳艺展中心及周边家居饰品商圈，广州的美林饰品中心便是后起之秀。

The largest home decoration market are located in Shenzhen Art Exhibition Center, Mayland Home Decor Center in Guangzhou is developing rapidly.

采购饰品学问非常深，虽然同一个位置，选择不同的饰品，效果会有所不同，但采购的时候一定要遵循美学原则，并考虑空间的需求，从而选择符合空间需求的产品。比如考虑数量、尺度、色彩、材质等因素时，都要综合考虑饰品与其他物品的关系，不能单看觉得好看就买了。设计师对一个优秀的软装空间作品的要求不亚于一个导演对一个演出片场的要求，任何一丝一毫都不能马虎，都有高要求，最后呈现的才是一个完美的软装空间作品。

Purchasing decorations has much knowledge. Even in the same space, different decorations make different effect. So when we choose decorations, we must follow the aesthetic principle and meet the requirement of space, such as the number, size, color, material etc. And we should consider the relationship between the decorations and other home items. A good soft decoration space it requires as much as a director works on a show. A perfect soft decoration work should be perfect in every details.

Confiture

## 七、采购制作环节的总结

在整个软装工程中，可以说第一阶段是停留在思想及纸面上的，而采购与制作作为第二阶段才是真正考验设计师的动手能力及组织能力的。一套好的方案，没有好的采购来配合，最后的成果可能会完全走样。

采购并不是按照清单照葫芦画瓢的工作过程，而是要求设计师不断地在设计方案与实际空间之间仔细思考，找到合适的切入点，这就要求设计师具备很强的设计能力、优秀的空间理解能力、对软装市场现有产品的清晰把握等多方面的综合素质。在软装公司中，具备采购能力的一般都是组长级别以上的设计师。

## VII. Summarizes Of The Procurement And Production

Throughout the soft decoration project, the first stage is to stay in the ideas and paper. The procurement and production, as the second stage, is the real test of ability and organizational skills of the designer. A good program can not be perfect without cooperation of great procurement, the final results will be completely.

Procurement is not to copy the list of models only, but needs designer to find the right entry point of design and actual space, which requires a designer with a strong design capability, excellent ability to understand space, and a clear grasp of the market for existing products and other aspects of abilities. In the soft decoration company, the person who has the ability of procurement can be able to be a leader or a manager.

软装行业的采购和其他行业的采购有很大的不同。有些人把软装归类为时尚行业和艺术行业，这其实是非常贴切的，因为从做方案到采购再到摆场，就是不断的艺术再加工的过程。方案、采购、摆场都是手段，并且各个阶段有各个阶段无法逾越的限制，比如方案阶段无法像纯艺术家一样天马行空地施展，总是要结合硬装及软装物品的可实现性来做方案，并且也无法做到所有的产品都是原创的，借鉴或组合反而是非常好的一种表达方式。

而采购阶段要结合方案和实际空间的情况，因为软装物品特别是饰品种类过于繁多，实际采购时经常会碰到更适合项目的软装物品，是忠于设计图纸还是忠于效果本身其实是很难选择的，但是谁又能拒绝更好的效果呢？所以一般碰到更加适合的软装物品时，有经验的设计师会用艺术家的心境忠于自己的内心感受，这已经是整个软装工程的二次创作了。

The procurement of the soft decoration industry differs from other industries. It is very appropriate that someone think it is part of fashion and arts industry, because from the design of program to purchase, then to the site furnishing, in fact, is the constant Art reprocessing process. Design, procurement, and site furnishing are the ways to complete project, and there are various stages of insurmountable restrictions, such designs can not be the same as the pure artist who can totally unrestrained display, soft designer has to combine hard mounted and soft decorations with the realizable designs, and every product used can not be all original, instead, copy or combination will be a very good way to express.

We have to in connection with design and the actual space to purchase items, because the soft decorations in particular are too many types, we will encounter more suitable items in actual procurement. To be loyal to the design plan or the final effect is actually a difficult choice, but who can refuse a better effect? In this circumstance when you encounter more suitable items for general , experienced designers will be loyal to their own feelings, which is already the second creation of the entire soft decoration project.

软装物品种类繁多，稍有不慎就会造成很多不必要的损失，在运输到项目地之前，要把入库及出库的工作做好。入库和出库工作的难易与软装公司的规模有关，但这个步骤一定要细心。

There is a wide range of soft decorations, the slightest mistake will cause a lot of unnecessary losses. we should be very careful on storage and delivery. Regardless of the scales, just pay attention on it.

**1. 入库注意事项**

(1) 检查所有收到的货品的完整性及破损情况，如有破损要及时联系供货方替换，因产品多而出现遗漏的情况也时有发生，所以应有专人检查。

1. The Precautions Of Storage

(1) Check the integrity and damaged situation of all received goods, if , contact the supplier to replace in time. And the situation of product omission has been occurred often, so we need a special person to check.

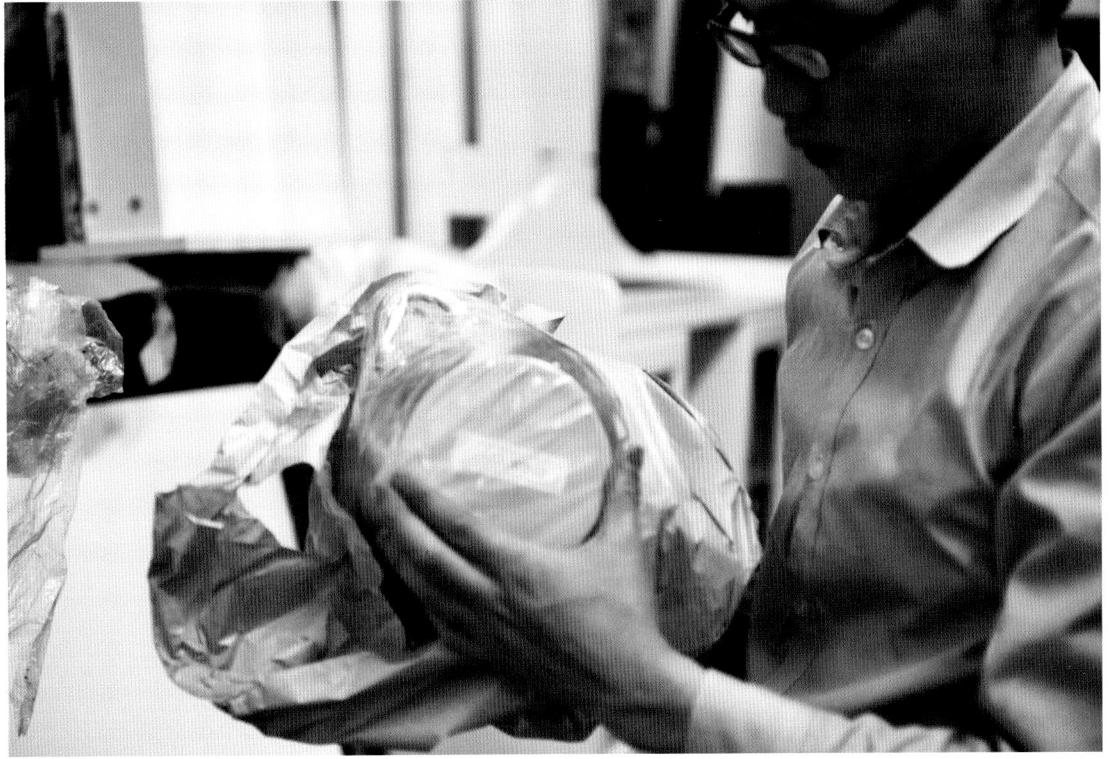

(2) 收到货品后要按要求摆放好，有很多易碎但又很贵的摆设品，稍不留神就可能碰倒、损坏。

(2) After receiving, the goods should be well placed according to the requirements. There are many fragile but expensive furnishings, which are too easy to be broken.

(3) 型号及户型要标注清楚，将编码贴好，外包装标明具体房型及空间，以免混淆。

(3) Model number and house type should be clearly marked, preferably properly labeled by code. All the information should be clearly marked on the package.

**2. 出库及运输的注意事项**

(1) 货物要清点清楚，不要有遗漏，二次发货的费用是非常高的。

(2) 尽量包装仔细，不要嫌麻烦，特别是饰品部分，基本都是怕压的，所以一定要包装好，不但里面要多包几层，外面还要尽量打木架，防止挤压。

2. Precautions Of Delivery And Transportation

(1) Check the goods and do not miss anything, the second delivery costs very much.

(2) Try to pack carefully, especially the decorations, most of which are fragile. They must be packed with more layers inside and gallows outside.

ITEM NO.
QTY. PCS
N.W. KGS
G.W. KGS
MEAS. 64 X 26 X 27 CM

(3) 按物品重量及抗压程度装车，重的、不怕压的放下面，轻的、怕压的放上面。

(4) 尽量专车专送，虽然现在的物流公司也很方便，但是只有尽量减少搬运次数才能最好地保护产品。因搬运次数多或搬运工人不小心造成的损失是极其大的，极其昂贵的家具，稍有不慎可能就会摔坏、跌坏、磨损、撞坏，造成很大的浪费。一般在装车及卸车前，一定要向搬运工人说清楚搬运的注意事项，可能只是多几句话，物品就可以避免几千元甚至上万元的损失费。

(3) Loading according to the weight and anti-pressure abilities of the items, put the heavy items below the light ones on the top.

(4) Deliver the goods by special cars. Although it is also very convenient to transport by logistics companies, only to minimize the number of delivery times can protect the goods best. The loss caused by rough handling are very large. Be much more careful to those extremely expensive furniture. Before loading and unloading, we must take the porters all together and tell them clearly about the handling precautions. Only few words can help to avoid millions of damages.

MADE IN CHINA

【微波电器产品】

# 陈设展示
# Furnishings and Presentation

## 1. 摆场步骤

(1) 做好现场保护，鞋套、纸皮等提前准备好。
(2) 摆好家具。
(3) 挂灯饰、窗帘、画。
(4) 摆设地毯。
(5) 摆设床品、抱枕、饰品、花艺。
(6) 细微调整。

## 1. Steps of Site Furnishing

(1) site protection, shoe covers and paper should be prepared in advance, etc.
(2) Arranging furniture.
(3) Hanging lamps, curtains, paintings.
(4) Putting carpet.
(5) Setting bedding, pillows, decorations, floral.
(6) Fine adjustment.

进场时要注意硬装环境，有时会遇到硬装未完成的情况，要及时与甲方、施工方负责人沟通，以免因不必要的麻烦延误工期。另外，应按照设计要求，将零部件直接摆放到具体的位置，有利于空间利用，也方便组装家具。

Pay attention to the hard mounted when entering into the room, sometimes we will meet the incomplete project, in this case, we should communicate with Party A and construction side at once, so as to avoid unnecessary trouble which cause the delay of the construction period. In accordance with the design requirements, placed directly to a specific location, not only in favor of space utilization, but also easy to assemble furniture.

在搬运带木框的货品时，为防止地板被刮花，应在进场之前在地面铺垫好纸板或保护物。家具拆包时要讲究方法，以免造成货物断裂或被割破。

To prevent scratching the floor when moving goods with a wooden frame, before entering the floor should be covered with good cardboard or protective material. And there are some methods to unpack furniture to avoid damages.

虽然有设计方案，但设计师在陈设现场通常会根据不同情况临时发挥，以达到最好的效果。

The designer may change the plan under different circumstances to achieve the best the results.

**2. 摆场各阶段的注意事项**

一定要保护好现场。硬装经过一个月或几个月的辛苦劳作，交给软装公司时一定要保护他人的劳动成果，不管是墙面还是地面、门、楼道，搬运物品的时候都要格外小心。在外出项目时，尽量找当地的搬家公司，他们因为经常搬运懂得很多该注意的地方，不管是效率还是对现场保护都会做得比较到位。

摆设家具时一定要做到一步到位，先把位置完全确定好，特别是一些组装家具，在这个房间装好后，要挪到另外一个房间时常需要经过拆卸才能进行，过多的拆装都会对家具造成一些损坏。

家具摆放好了后，就可以确定好挂画和灯的准确位置了。如果没有装好家具就挂画或挂灯，很容易把位置弄错，而一旦修改就会对硬装部分造成一些损坏，所以顺序不能颠倒。挂画对位置和高度都很有讲究，有些设计师不太讲究，结果挂得一高一低或者稍有歪斜，整个空间的品位就会大打折扣了。专业的挂画师傅是要带水平仪的。

这里具体说说挂画、安装灯、安装窗帘时应注意的细节。

2. The Precautions of the Site Furnishings

We must make sure to protect the site carefully. After a few months of hard work, the hard mounted comes to the hands of our soft decoration companies, we have to respect the labor of others. No matter walls, floors, doors, or corridor, we must be very careful when handling. And try to find a local transport company, because they have a lot of carry experience which are good for efficiency and on-site protection.

Site furnishings must be done in one step. First make sure of the location. Some assembling furniture has to be demolished to complete in one room and placed in another one. Too much dismounting and installation will cause some damage to the furniture.

After the installation of furniture, the exact location of paintings and lamps can be determined. If hanging the paintings or lights first, it is easy to be putting the wrong positions. The installed modifications will result in some damage of hard mounted, so the order can not be reversed. In fact, the position and height of handing paintings are very particular, which some designers do not pay attention to, the result of one high and one low, or slightly askew, impact greatly on the entire space effect. So professional workers will use level instrument when they work.

Some details we should pay attention to when we are hanging painting, installing lightings and installing the curtain.

**(1) 挂画时的注意细节**

①工程部人员挂画时要与设计师沟通好具体位置，如多幅画在一起时每幅画间隔的距离等细节。

②挂画要先用尺子量出挂画两边挂钩位置正确的距离，用冲击钻在墙上打出螺丝孔，不能直接在墙面用水泥钉等直接钉钉子，因为这样容易损坏墙面。

③挂画应与地面水平，不能出现高低不齐的情况，挂好后及时用水平仪测量挂画是否水平。

④挂画高度对视觉有很大影响。人的正常水平视觉，其视线范围是在上下60度的圆锥体之内。所以，最合适挂画的高度是离地1.5~2m的墙面位置（设计师有特殊要求的除外）。墙上画面应该向地面微倾，且不能看到钉子和钉孔。放置于柜子上低于人眼的画面，则应以仰倾向顶棚的角度摆置，以方便观赏画面。

⑤遇到大幅挂画不打钉的情况，要先和甲方沟通清楚，用玻璃胶或AB胶涂抹在挂画后面，摆好位置后用胶带固定好，一天后胶干即可。大画一定要挂结实，安全问题至关重要。

(1) Pay Attention to Details When Hanging Paintings

①When hanging the paintings, designers should communicate with workers to determine the position and the height of paintings. If there are many paintings, the interval distance of each painting should be noted.

②When hanging the paintings, firstly, measuring out the correct distance of the hooks, madding screw holes with the impact of drilling, instead nailing in the wall which may be contaminate the wall.

③The ground level should be considered when hanging paintings. After hanging up, using level measurement to maintain the level with the ground.

④It has great influence on the vision with the height of the hanging paintings, one's normal level of vision is in the upper and lower within sixty degree cone. Therefore, the most appropriate height of paintings is from 1.5 to 2 meters on the wall (Except for the special requirements from designers). The picture should be leaning to the ground and can't be see the nail and nail hole. The paintings should be placed supine tendency smallpox which be put on the inconspicuous cabinet, convenient for enjoying.

⑤We must communicate with Party A clearly when using no nailing, and should using glass glue or AB glue smearing behind the paintings, adhesive tape the picture after placing. It is completed after the glue dried. Be sure to hang large paintings firmly. The security issues is the most important.

**(2) 安装灯时的注意细节**

灯在每个项目里都是点睛之笔，在需要安装时要格外地注意安全，确保安装期间和安装好后不会发生危险。

①安装灯之前要先阅读灯具的说明书、熟悉电路等，对灯具的配件要检查清楚，查看是否有损坏。

②因为水晶灯都非常重，预先打好的膨胀螺丝至关重要，螺丝的大小要与安装的灯具大小重量成正比。

③灯具里的接线要在灯具的底壳内，不能让电线裸露在外，灯头的绝缘外壳不能有破损和漏电，装白炽灯的吸顶灯具，灯泡不能紧贴灯罩，否则容易发生危险。

④灯具安装好后要检查所有灯泡能否正常开关。不锈钢、水晶灯具安装要戴手套。安装后要擦拭干净。

(2) Pay Attention To The Details Of The Installation Of Lighting

The lighting is the key point in every project. When we install lighting we have to be very careful to ensure no dangers during and after installation.

①Read instructions before installing the lighting, have an intimate knowledge of electric circuit, check accessories of the lighting, and find out if there is any problem.

②Because crystal lamps are very heavy, Setscrew which be sunk beforehand are crucial, the size of the screw has to be proportional to the lamps' sizes and weights.

③Lamp wires should be put into the bottom of the lamp body , and can't be exposed outside, the gas-insulated shell of lamp holder can't be damaged and leaked, incandescent lamp and ceiling lamp can't be closed to the lampshade, or it may cause danger.

④We should make sure that all light switches working properly after the installation. When installing the stainless steel and crystal lamp, we should wear gloves, and clean up the lamps after installation.

**(3) 安装窗帘时的注意细节**

①窗帘挂上去后要调试一下，看能否拉合，高度是否合适，如果还需进一步清洁卫生，就需要先把窗帘用大的塑料袋保护好。

②窗帘杆或轨道应固定在承重墙上以确保稳固，在轨道每隔20cm处应打固定孔，这样才能保证不会发生坠落事件。

③安装窗帘轨道要在窗帘盒的中心定位，轨道要直，尤其是宽大的落地窗，不能仅靠目测。找平后检查窗帘轨道的预埋配件位置和方式是否符合要求，如有平度、出墙距离等误差应及时处理。

④窗帘的下沿应低于窗台10cm左右，落地窗窗帘的距离一般在3~5cm。窗帘过长容易弄脏并给人拖拉感，太短了悬在地面上影响整体美观。

(3) Pay Attention to Details When Installing the Curtain

①We have to make a slight adjustment after hanging the curtains, to check if the curtains can be pulled together, and whether the height is appropriate. If it needs to clean the room before, it had better to further protect the curtains with a large plastic bag. As shown in the following:

②In order to guarantee the stable and not to fall down, the curtain rod or the track should be fixed in the bearing wall; it should make the fixing hole every 20cm in the track.

③When installing the curtain rail, we should put it in the center of the curtain box. The rail must be straight, especially the Large French windows. After leveling the embedded parts examination curtain track position and the way it meets the requirements, such as roughness, wall distance errors should be timely treatment.

④The curtain should be below the windowsill about 10 cm, the distance of the window curtains should be about 3-5 cm. If the curtain is too long, it gets dirty very easily and give a person the sense of procrastination, on contrast, it will hang on the ground and affect the overall appearance.

**(4) 其他**

一个卧室中非常重要的部分就是床品，虽然材质好、颜色好，但是如果摆设不好也是没用的。该叠好的叠好，该拉直的拉直、铺平，棉芯均匀，抱枕饱满，摆放讲究，这样做出来的感觉就显得非常有生机、有朝气。

饰品部分根据实际情况摆设，只要效果好，位置可以适当调整和互换，注意整体的把控就可以了。

(4) Others

Bedding is a very important part for bedroom. Without proper placing it will be useless even if we have good materials and best color of items. Folded, straightened, paved, cotton core uniform, plump pillows, placing in good way, all those will bring a feeling of vitality, and energetic.

The placing of decorations will depend on the real situation. As long as the effect is good, the location can be changed.

CHAPTER THREE

## 汉意堂案例展示

## HANITON CASES

# LEISURE AMERICAN STYLE CREATING A UNIQUE ARTISTIC CHARM

## 休闲美式 写就别样艺术韵味

这处被设计师赋予宁静和浪漫的空间被定义为休闲美式风格，意在构建一个安静、自然、舒适、温馨的休闲生活情趣空间。

This case is a quiet and romantic and leisure American style, trying to create a peaceful, natural, comfortable, warm and casual living space.

## 家具配置

## FURNITURE CONFIGURATION

在家具的选择上，设计师不仅特别注重其功能性和舒适性，还注重其优雅、大气的外观。纵观室内，不加修饰的原木家具散落在空间的各个角落，从墙角的餐吧柜到客厅的电视柜无处不在，像这个家的主人在诉说着渴望过一种与自然接轨的生活。平实却不失舒适的沙发，给人带来淡雅、清新的感觉，呼应了设计的主题。在饰品选择上，主要以丰富的内容与美丽的画面效果烘托环境，或是家禽，或是器皿，抽象的、具象的不一而足，皆以自身独特的内涵与情境带给人丰富的感官体验，呈现出虚实、动静结合的美感，为休闲的居家环境增添了一丝艺术气息。在空间线条的处理上，强调“拱形”结构的灵活运用，营造强烈的空间层次感，同时赋予空间浪漫、梦幻的氛围。

Designers paid attention to not only function and comfort, but also appearance when chosing the furniture. Throughout the rooms, the plain log furniture scattered in every corner, and just like the owner is telling the desire to live a life in line with nature. Simple but comfortable sofas, brings elegant, fresh feeling, echoing the design of the theme; In the decorations selection, mainly using some items which are rich of themes and pictures to dramatize the environment, like poultry or household utensils, some are abstract, some are representational, but all of them have their own unique meaning and context to bring people rich sensory experience, showing the beauty of fantasy and fact, and the static and dynamic combination to add some artistic taste to this leisure home environment. In the processing of space lines, using of "arch" structure, create a strong sense of space levels, making the space romantic and dreamlike.

EATING LOCAL
EATING LOCAL
marie claire
luscious

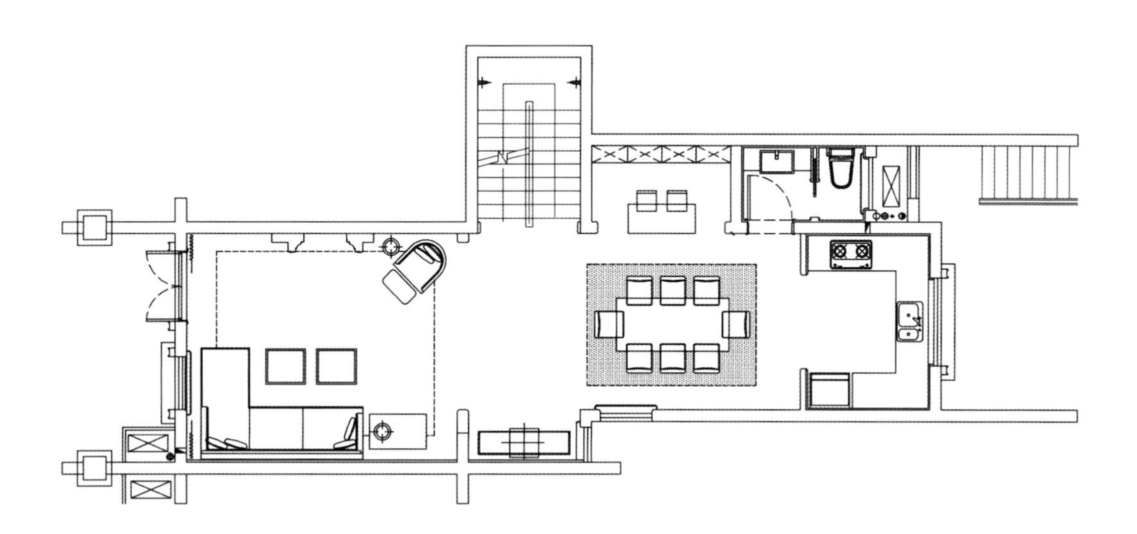

一层平面图 1st Floor Plan

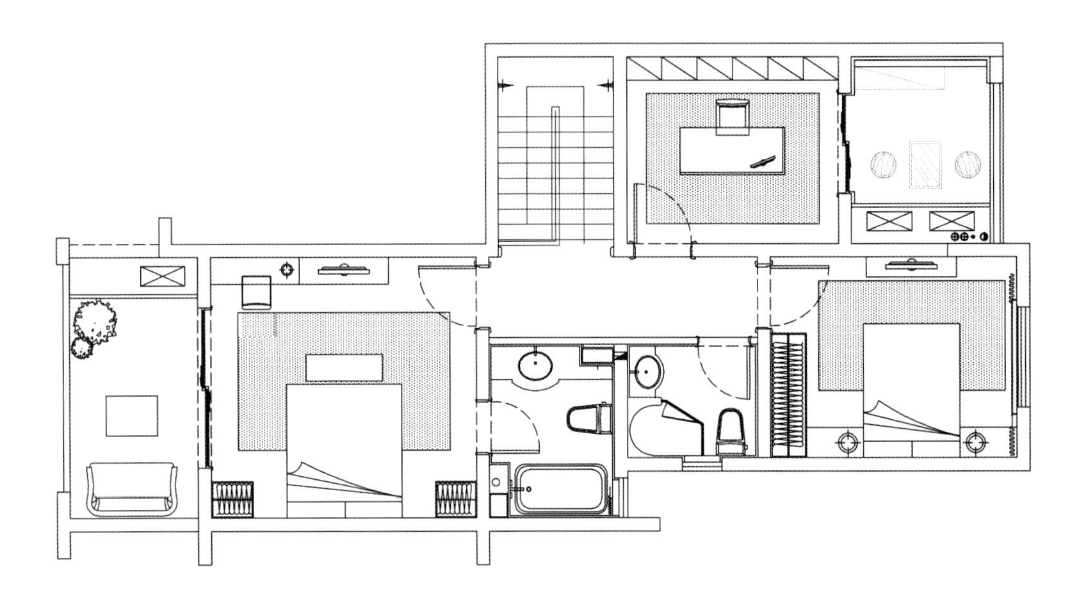

二层平面图 2nd Floor Plan

## 色彩搭配

COLOUR COLLOCATION

空间中明快的色彩不加雕琢，彰显纯净的美。尤其是一层客厅以大自然中常见的色彩铺陈空间，顶棚、地板的木色搭配沙发的米白色及椅子的绿色，给人一种清新之感。其他区域延续了一层的格调，彰显出无处不在的自然与温馨。除此之外，印有花卉图案的壁纸是这个空间的一大亮点，布艺、壁纸侧重于对花卉图案的呈现。设计师大量运用艺术类的装饰画作为点缀。用以美化环境的清新、自然的花卉绿植在室内随处可见，使得整个家居空间更富自然的色彩与温馨的气息，彰显业主自身的生活态度和追求。

The bright and pure colors without embellishment bring a kind of fresh feeling of nature. The main color of living room in the first floor is common but comfortable. Wooden ceilings and floors, creamy-white sofa and green chairs match each other properly. Other are the same style, natural and warm. In addition, wallpaper with printed flowers highlights the whole space. Decorative flowers and plants can be seen everywhere in the house, showing how much the owner loves the life.

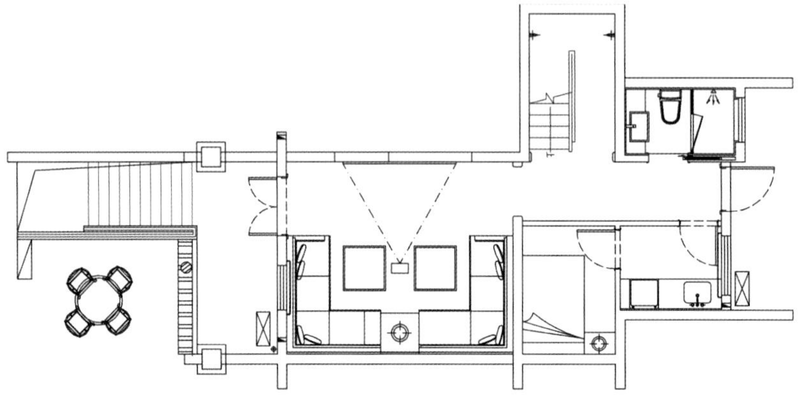

负一层平面图 -1st Floor Plan

本案整体空间清新淡雅，无处不在诉说着都市里少有的一种休闲与自然的生活方式。

In this case the overall space shows a fresh and elegant atmosphere, and tells a casual and natural way of life, which is rare to find in the city now.

# LEISURE AMERICAN STYLE CREATING DISTINCTIVE COMFORT

## 休闲美式 独具一格的惬意

本案融合了不同风格的优秀元素，形成了独具一格的休闲美式风格。以舒适机能为导向，摒弃以往的烦琐与奢华，强调了自然的重要性，重点突出了生活的舒适和自由。

This case combines outstanding elements of different styles, forming a unique leisure American style. It is a comfortable house without tedious and luxury decorations, showing the importance of nature, and highlighting the comfort and freedom of life.

## 家具配置

## FURNITURE CONFIGURATION

本案所配置的家具材质多样，其中皮革、布面与木材的结合最为常见，使家具更为舒适，同时彰显低调、内敛的气质。部分家具可见雕刻点缀的花纹，未经加工的质感，赋予空间自然、朴实的乡村气息。其他用具选择舒适、实用和多功能的，不过分强调繁复的雕刻和细节，营造返璞归真的境界，符合现代人表达自由与个性的诉求。 整个居所内，如此多的实木家具、铜质饰品及完美的色彩搭配，怎能让人不流连于如此悠然自得的家。

The materials of furniture are various in this case. Leather, cloth and wood are the most common, making the furniture comfortable and functional with elegant feature. Some furniture, beautifully carved, looks rough and old, bring us into a rural life. Then other appliances are useful and multi-functional, simple in appearance. So much wooden furniture, copper decorations and perfect colors are totally matched with modern life. Who are willing to refuse to stay in such a lovely home?

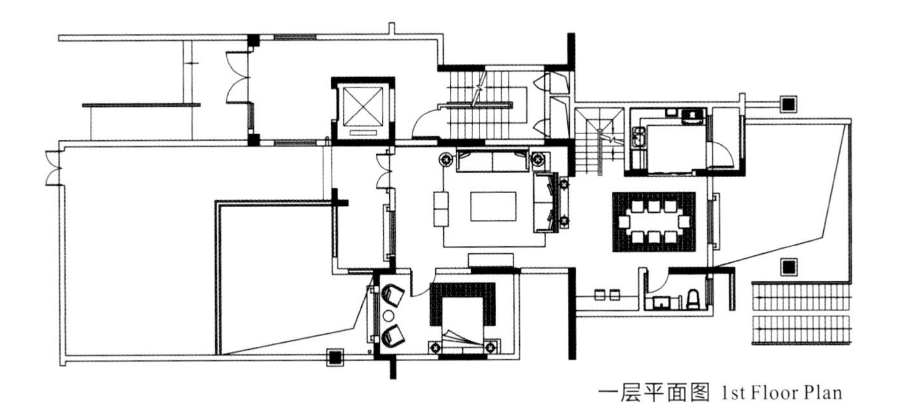

一层平面图 1st Floor Plan

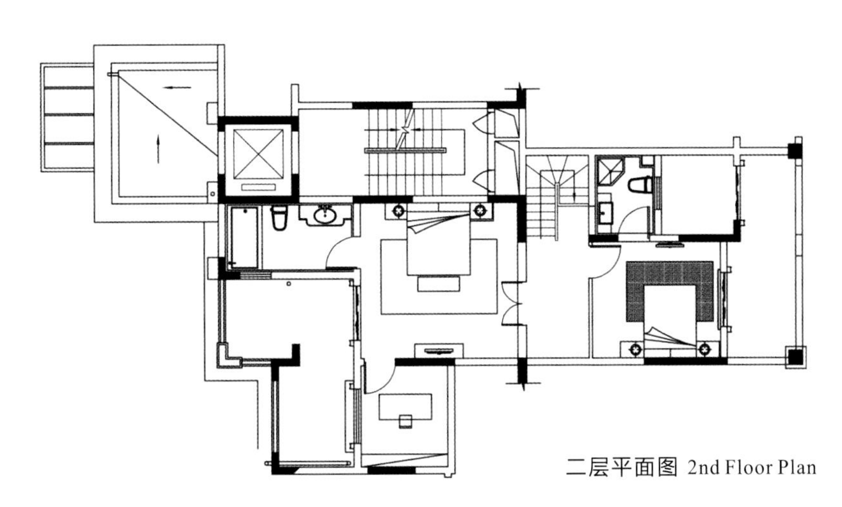

二层平面图 2nd Floor Plan

ITALIE
ITALIE
ITALIE

TABLEAU
TABLEAU
ENCYCLOPEDIQUE
ET METHODIQUE

## 色彩搭配

## COLOUR COLLOCATION

空间主要以褐色、红色、绿色、米白色等自然色调为主，顶棚及地板以大面积的木色铺设。为了呼应这种家居色彩，设计师在软装上透过一些空间和家具的留白来配和。大部分家具以深色系为主，再混合窗帘、地毯、布艺、装饰画的色彩，加以饰品花艺的点缀与调配，舒缓了视觉效果，使空间出现了层次的变化，带来一丝灵动的气韵，提升了空间的品位与质感。多功能厅的色彩搭配最为丰富，借由窗帘、床品、地毯的色彩混合和过渡，呈现出丰富的视觉效果。

Space is mainly in brown, red, green, beige and other natural color tones, Ceiling and floor are wood color. In order to keep balance, designers leave some blank in soft decoration. Most of the furniture is dominated by dark tones, but with the help of colors of curtains, carpets, cloth and decorative paintings, the whole space looks vivid and elegant. Especially in the multi-functional room, various colors in curtains, bedding and carpets match so well that the visual effects are perfect.

本案房型大气、功能丰富，从软装设计到布料选装，每个细部都充满着美感。通过美式风格来彰显别墅的典雅气质及休闲、惬意的生活品质。多功能厅展示了业主的热情好客，在这里可以随时与朋友品酒，分享生活的休闲与惬意。

This Case, with a great house type and rich functions, is beautiful in every detail from soft-mounted design to fabric option. American style highlights the elegant, leisure and comfortable life. The multi-functional living room shows the hospitality of owners. Here you can feel free to taste wine, and share casual and comfortable life with friends.

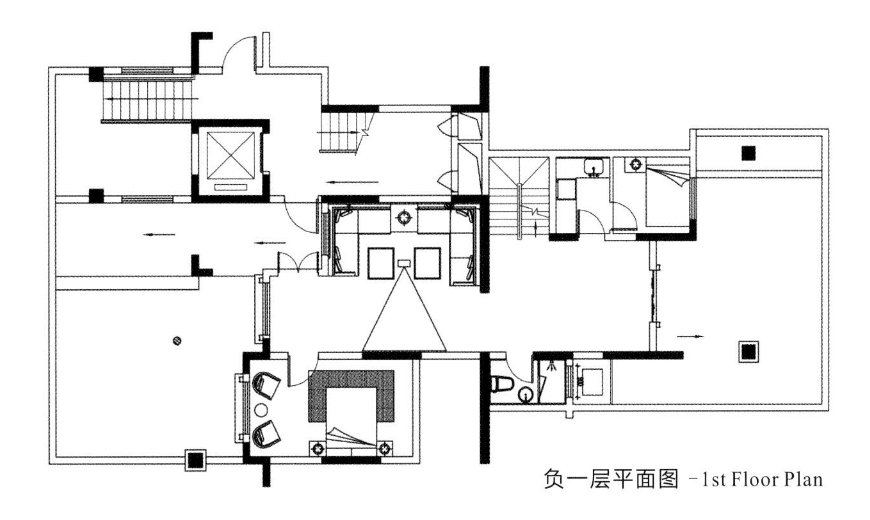

负一层平面图 –1st Floor Plan

# A QUIET DREAM
# A SHADOW OF LOVE

## 一帘幽梦 一影落情

每个人都想拥有一处生活的理想居，无关金钱和地位，只寻求休闲与品质、简约与乐趣。该案中设计师以木色为主调，搭配具有温馨浪漫气质的布艺等材质，加之主题性艺术陈设的点缀，使主人的爱好与生活品位淋漓展现。这样精致的空间维度折射出的是业主浪漫、休闲的都市生活哲学。

Everyone wants to have a perfect home, not about money and status, only matters of leisure and quality, simplicity and fun. In this case designer use wood color as the main color, combined with a warm and romantic cloth and other materials, plus decorative theme of art furnishings, to show the owner's hobby and lifestyle. In this exquisite space, it reflects the romantic casual urban life philosophy of the owners.

## 家具配置

## FURNITURE CONFIGURATION

为了呼应主题风格，设计师采用了美式风格的家具，其身上特有的自然、经典、斑驳老旧的印记，似乎能让时光倒流，让生活慢下来。无论是客厅、卧室还是餐厅，家具的线条都简洁明晰，外加带有沧桑岁月感的配饰和得体的装饰，营造出闲散与自在，温情与柔软的氛围，给人一个真正温暖的家。

Designers use the American-style furniture. Its unique natural, classic, mottled old mark seems that it could turn back time, to make life slow down. Whether in the living room, bedroom or dining room, furniture are all simple and accessories are appropriately decorated with the vicissitudes of the years, creating an idleness and ease, warm and soft atmosphere, giving a truly warm home.

## 色彩搭配

## COLOUR COLLOCATION

木色是这个空间的主色调，或深或浅，搭配适度。色调搭配层次分明，跳跃的蓝色和花布悄悄地置入其中，自然中融入浪漫的色彩和元素，使得空间立刻明朗起来。卧室布置较为温馨，作为客户的私密空间，软装用色非常统一，以功能性和实用、舒适为考虑的重点，用温馨、淡雅的成套布艺来装点，使整个居室在浓浓古韵中渗透出休闲、舒适的气息。

Wood color is the main color of this space, deep or shallow, with a moderate balance. A well-bedded combination of colors, mixing some outstanding blue and figured cloth to put some romantic elements into nature, making the space immediately clear up; the decoration of bedrooms are warm. As a private space, the use of colors are very uniform, functional, practical and comfortable, with a warm and elegant fabric to decorate the whole room penetrate the casual and comfortable atmosphere in the archaic Chinese rhymes.

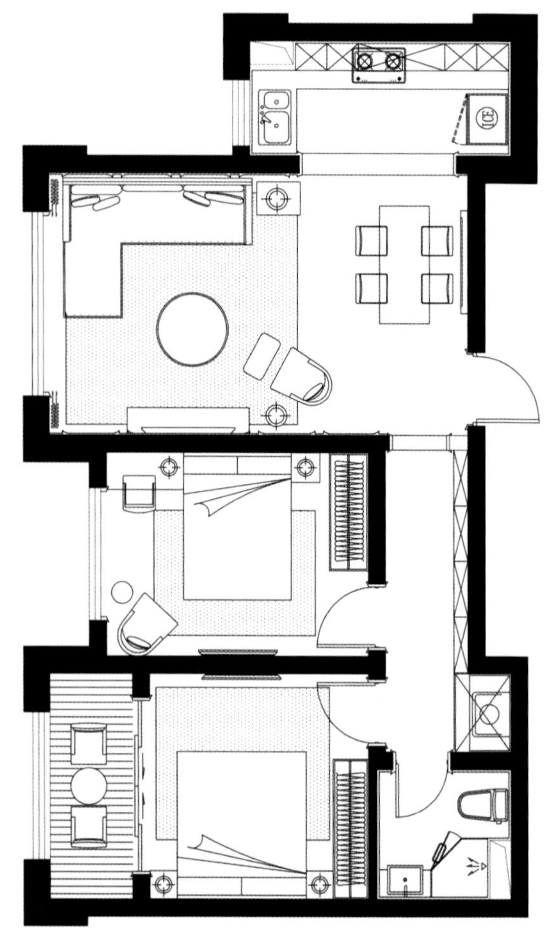

# ENJOY THE EXOTIC STYLE DREAM HOME

## 享受异域居家情调的梦想

本案是一套富于梦想的居家样板空间，以表现舒适和自由为主题。设计师在这里将梦想具象化，将业主的爱好和对具有个性的装饰品的喜爱融入空间。蓝色的主调搭配着业主在旅游途中从各地带回来的纪念品，尽显质朴与华美。如此温馨的异域居家情调，让业主充分享受梦想的乐趣和美好。

This case shows a house full of dream. Comfort and freedom are the theme of this space. Here designers visualize the dream of owners by adding their favorite ornaments into the space. The blue tone of the space mixes with the tourist souvenirs which are brought back from everywhere by the owners, showing a plain gorgeous of the space. Such a home with warm and exotic atmosphere let the owners fully enjoy the pleasure and beauty of dream.

## 家具配置

### FURNITURE CONFIGURATION

在家具配置上主要考虑两点：审美品位和舒适。

首先，为表现舒适和自由的主题，使用的家具都带有浓烈的大自然气息。简单的线条、粗犷的体积、自然的材质，以及带有时间感的配饰，充分彰显自然与舒适，让业主内心始终充满对自由的无限向往，渴望自由、随意的生活。

其次，为了符合业主的审美品位，保留原木色的家具，为家增添了一种质朴之感。客厅是客户踏入样板房的第一驻足点，这个空间必须给客户“这就是我想要的家”的感觉。为了达到这一效果，设计师选用了造型圆润的三人沙发，让客厅饱满而丰富。餐厅中有六人位的方桌，方桌的使用使整个餐厅更显宽敞。精致的吊灯散发出温馨的光芒，精美的餐盘、餐具和花艺摆件透露着业主精致的生活。不同布艺的两张餐椅，打破了以往餐椅全部用同一种颜色的呆板印象，能给客户带来小小的惊喜，无形中也对开发商增加了好的印象分。

In the furniture selection, there are two main aspects are taken into consideration, aesthetic taste and comfort.

The first is to express the theme of the comfort and freedom, using the furniture with a strong natural flavor, simple lines, rough volume, natural materials, and accessories with a sense of history, to achieve a natural, comfortable purposes. Let the owner's heart always be filled with the desire for the free and casual life.

The second aspect is to meet the aesthetic taste of the owners. Wood color of furniture add a sense of rustic for the home. Living room is the first stop point when we enter the show house, and this space must make clients feel that “this is the home I want”. Therefore designers use sleek trio sofa to make the living room full and rich, and the 6 people square table making the whole restaurant more spacious. Exquisite chandelier exudes a warm light, delicate dinner plates, tableware and floral ornaments, revealing the exquisite lifestyle of owners.

## 色彩搭配

## COLOUR COLLOCATION

蓝色是这个空间的灵魂色，其所代表的梦想意义与业主对自由的憧憬十分吻合。但在空间中，设计师并没有大篇幅使用蓝色，而是将其巧妙地点缀在空间中，令其更为鲜明和突出，如餐厅的油画。从客厅到餐厅采用一脉相承的色彩搭配的手法，颇为引人注目的橘色与深沉的蓝色搭配，具有较大的跳跃性，将热情、明快与自然、朴素的感觉融为一体。

Blue is the main color of this space, which is the color of dreams that matches the color of freedom that the owners are longing for. Instead of painting it everywhere, designers use it wisely in proper places to make it more distinctive, such as the painting in the restaurant. The style of the dinning room follows the living room's. Bright orange mixed with deep blue shows both passion and calm.

整体空间中营造的是一种温馨的家居氛围，既有沉静的基调，也有斑斓的色彩。材质的运用使色彩更具美感。

The whole space express a warm feeling. It is peaceful and colorful.

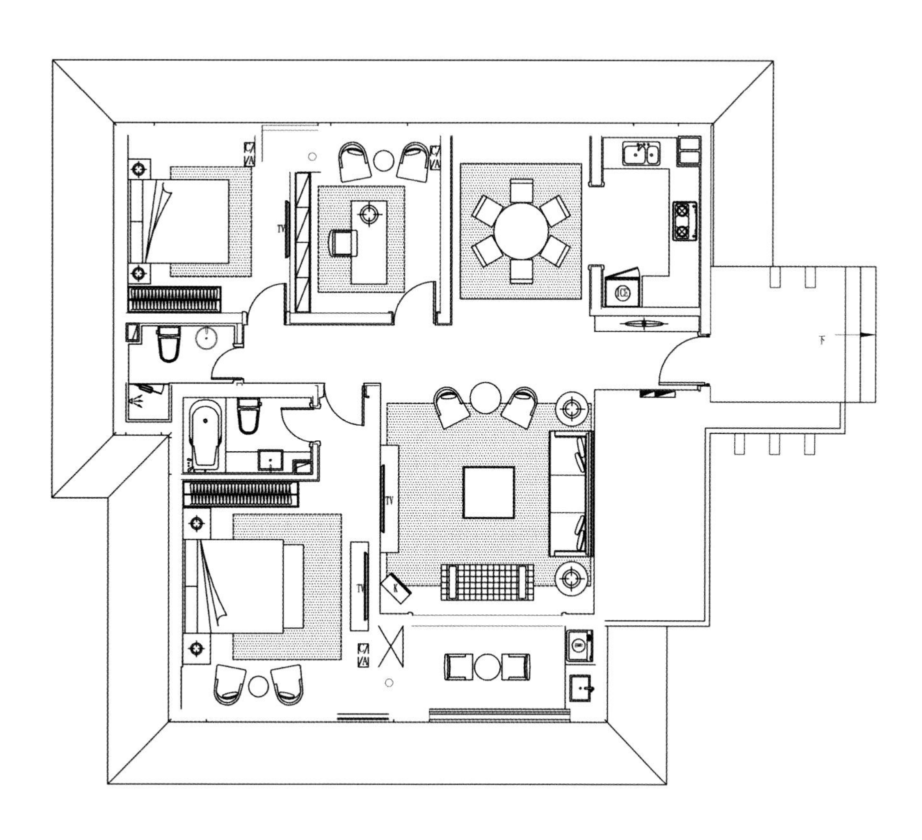

# Passion For Mediterranean Telling A Distant Dream

## 浓情地中海 述说悠远的梦

在硬装纯板式住宅的基础上，该案的软装设计非常人性化。设计师充分考量了业主居家生活的动线，空间设计动静结合，极具品质感。

The soft decoration makes this residential house very user-friendly. Designers fully considered the owners' hobby and made a perfect house with convenient moving line and comfortable rest area.

## 家具配置

**FURNITURE CONFIGURATION**

设计师尽可能保持原有硬装的轮廓，软装上注重体现业主的生活细节。首先，空间中没有使用过多的装饰物，简单的物件也能突出业主的品位，如卧室里活灵活现的动物摆件、书房中的老式电话机等。其次是客厅的窗帘和挂画，地中海风格一目了然。电视柜上的陶罐饰品、互相呼应的家具组合，组成了一幅恬淡的风景画。卧室的家具选择也沿袭客厅的风格，利用充满地中海气息的小物件和装饰画搭配，让整个卧室都充满了生机，并配有层次丰富的花艺作为点缀，使得卧室变得更加的明亮、宽敞之余还有温馨、舒适之感。

In this case, the soft decoration, focus on the details of life. First of all, do not use too many decoration in the space, while simple items can also highlight the owner's taste, such as vivid animal ornaments in bedroom, old-fashioned telephones in study room. Then, Mediterranean style are easy to see on the curtains and paintings of the living room, pottery decorations on the television cabinet, furniture combination echoes each other, forming a quiet and comfortable Mediterranean landscape scenery. The bedroom furniture selection also follows the living room features. The use of small Mediterranean-style items and decorative paintings make the whole bedroom full of vitality. Combined with the rich layers of floral embellishment, making the bedroom more bright and spacious, and bringing a warm and comfortable feeling.

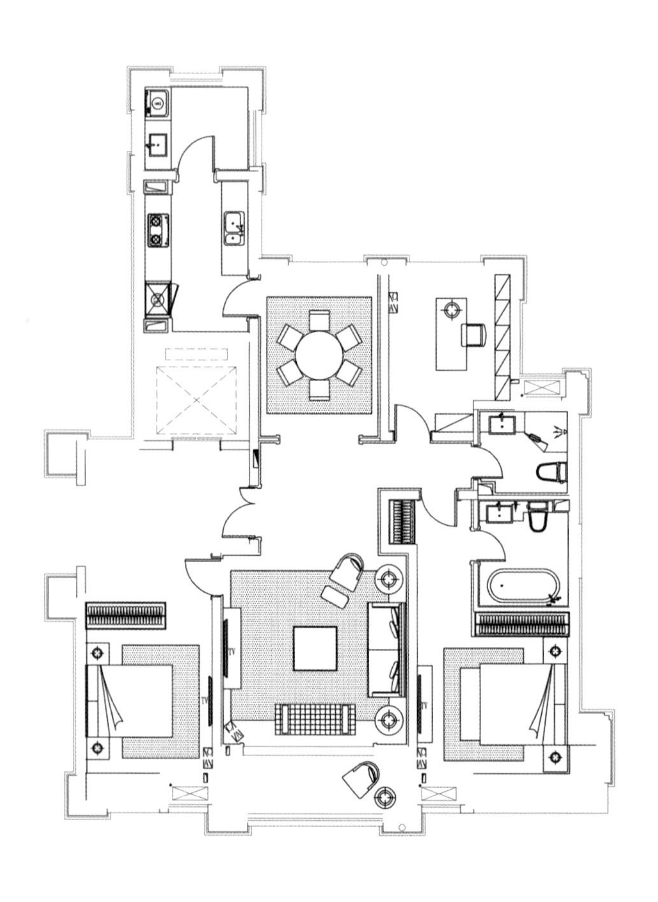

## 色彩搭配

## COLOUR COLLOCATION

客厅色彩的搭配十分协调，单色或多色的组合都不显得过于突兀，鲜红、青绿、米白，如此层次色调丰富的鲜花，争相斗艳地开放着，在客厅里显得特别抢眼。卧室中的饰品组合的色彩也比较丰富，雕花的抱枕、壁纸及窗帘颜色呈浅色，在用色上达到统一。而书房的做旧的铁艺吊灯无论是色彩还是款式都独具特色，让整个空间更加丰富多彩，又能把客户的视点往上方吸引，使空间在视觉上得到了延伸。如此浓郁情调的地中海气息，像是在向人们述说一个古老的故事。

The color of the living room is very harmonious and balanced, no matter monochrome or multicolor combinations. Red, green, white, such rich colors of flowers, vying with each other for glamour in the living room; the color combination of bedroom decorations is also abundant, light – colored pillows, wallpaper and curtains to achieve unity in the use of color. While the old wrought iron chandelier in the study room is unique in either color or style, making the entire space more colorful and can also draw the customer's attention to the top, letting the whole space to be extended. As if telling people an old story in Mediterranean.

# SEA AMBER HYTING THE MODEST DIGNITY

## 海珀香庭 低调的尊贵

该案位于西安高新区，是一套面积为230㎡的精装样板房住宅，由汉意堂担任后期的软装设计。根据原有的硬装，设计师考虑如何让典雅浪漫的情怀在软装中得以体现，于是决定把少量中式元素与法式古典设计相融合。走进内里，空间中散发出的是高贵、典雅、精致、浪漫的独特气质。

The case is a 230㎡residential show house, located in Xi'an Hi-tech district and the soft decoration is designed by Haniton. According to the original hard mounted, now designer should consider how to make elegant and romantic feelings be reflected in soft decoration, finally the designers decided to put some Chinese style element into the French classical designs. Walking inside, the space exudes a noble, elegant, refined, romantic unique atmosphere.

## 家具配置

## FURNITURE CONFIGURATION

空间的家具以法式为主，融入些许现代时尚的元素，混搭少数中式饰品作为点缀，使空间浪漫、尊贵又显内敛、稳重。材质上应用了实木贴皮、手绘、雕刻贴箔，整体打造具有西方形态、东方风情、优雅时尚感的空间，让东方美沉浸在西方华贵的细节之中。如客厅中的沙发及地毯，原本给人一种法式新古典的感觉，但设计师巧妙地使用了两幅孔雀的挂画，瞬间将整个客厅从西方风情拉回到了东方。更值得一提的是老人房的窗台装饰，一张椅子、一本书、一个鸟笼的摆件，这些手法看似简单，却在空间中营造了一个生活场景，突显出业主的品位。

The space mainly uses French style furniture, and put a little of modern fashion element, mix and match a few Chinese decorations, to make a noble and romantic space. For materials, the application of the wood veneer, hand-painted, carved foil sticker, overall creates a Western form with a oriental style and elegant fashion sense of space. Let oriental beauty immersed in the Western luxurious details, such as the sofa and carpet in living room, originally giving a French neo-classical feeling, but the designers use two peacock paintings, here, all of a sudden bring the entire living room from the West back to the East. It is worth mentioning that the windowsill decoration of elder room, using ornaments such as a chair, a book, and a birdcage, seems to be so simple, but creates a scene of life to highlight the owner's taste.

## 色彩搭配

## COLOUR COLLOCATION

该案的软装设计整体色调运用了低对比度的米色、蓝色、咖啡色，并搭配恰到好处的金色，与硬装的米白色调交融，呈现出来的空间既低调又不失尊贵与雅致。

色彩搭配在细节上的处理也很到位，镶嵌在绒毛抱枕上的白色珍珠、温馨的布艺搭配花鸟图壁纸，以绒布和丝为主的布艺，加上拼花地板的运用，别有一番风味。而饰品多以陶瓷和铜为主，自然、朴实的色调让人感受到更深的文化和历史感。

The soft decoration of this case mainly uses low-contrast colors, such as beige, blue, coffee color, and mixed with gold to make a noble and elegant space.

Colors on the details are also very exquisite. The white Pearls set on the white fluff pillows, warm fabric with a bird and flower wallpaper, the fabric with velvet and silk materials, the application of matched floor, all of those showing a good taste. While many decorations are made of ceramic and copper, the natural and simple colors makes people feel a deeper sense of culture and history.

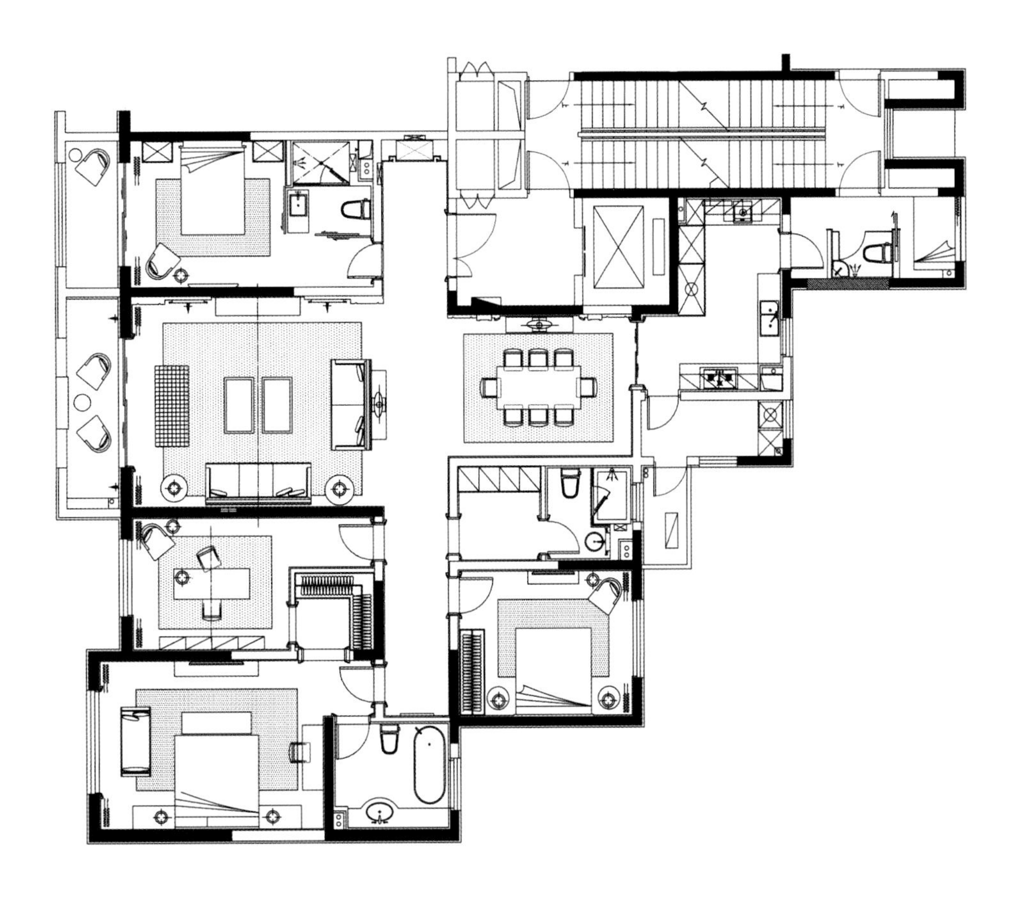

TINDRA
TINDRA

# THE ROMANTIC QUALITY SPACE

## 散发浪漫情怀的品质空间

本案汲取法国艺术的精华，以法式浪漫情怀为基础，还原一处具有浓郁巴黎气息的高品质生活空间，为业主带去浪漫优雅的艺术生活体验。在空间布局上，设计师强调整体协调关系，在设置足够的储藏收纳空间的同时，让其他功能的区域空间尽可能地放大。

This case derives the essence of French Art, and bases on romantic French style, to create a quality living space with a rich flavor of Paris, and bring the romantic and elegant art of living experience for the owners. On space layout, designers emphasize the relationship of the whole space, while maintaining adequate storage space at the same time to enlarge the other functional areas of space.

非交付标准

## 家具配置

**FURNITURE CONFIGURATION**

在家具的搭配上，设计师更多地从空间功能需求上来选择。客厅是接待客人和家人活动的主要场所，没有过多的装饰。两组沙发与单人椅的搭配让客厅饱满、大气且不失层次感。同时又恰到好处地置入一些现代感强又非常时尚的家具，给居家生活增添了活泼的气氛。主卧的配饰不同于客厅，流畅的家具线条和唯美的造型，让每个身临其境的人都能感觉到女性的柔美与雅致，更为每个细节的巧妙设计所惊叹。

For the collocation of furniture, designers select the furniture more depend on the space functional requirements. The living room is the main place for the reception of guests and family activities. There are not too much decorations here. Two sets of sofas together with single chair creating a full space and giving a space hierarchy feeling, at the same time, placing some strong and very stylish modern furniture adds to home life a sense of lively and cheerful. The decoration of the master bedroom is different from the living room. Smooth furniture lines and elegant styling, let people who enter this beautiful and elegant space be amazed by every detail of the design ingenuity.

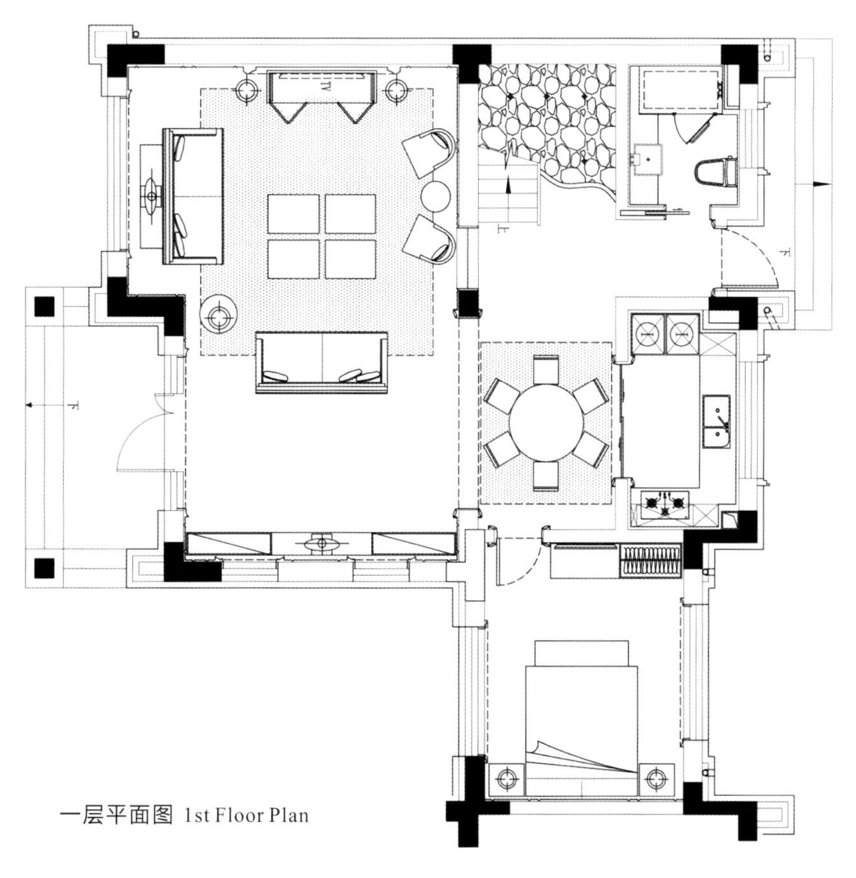

一层平面图 1st Floor Plan

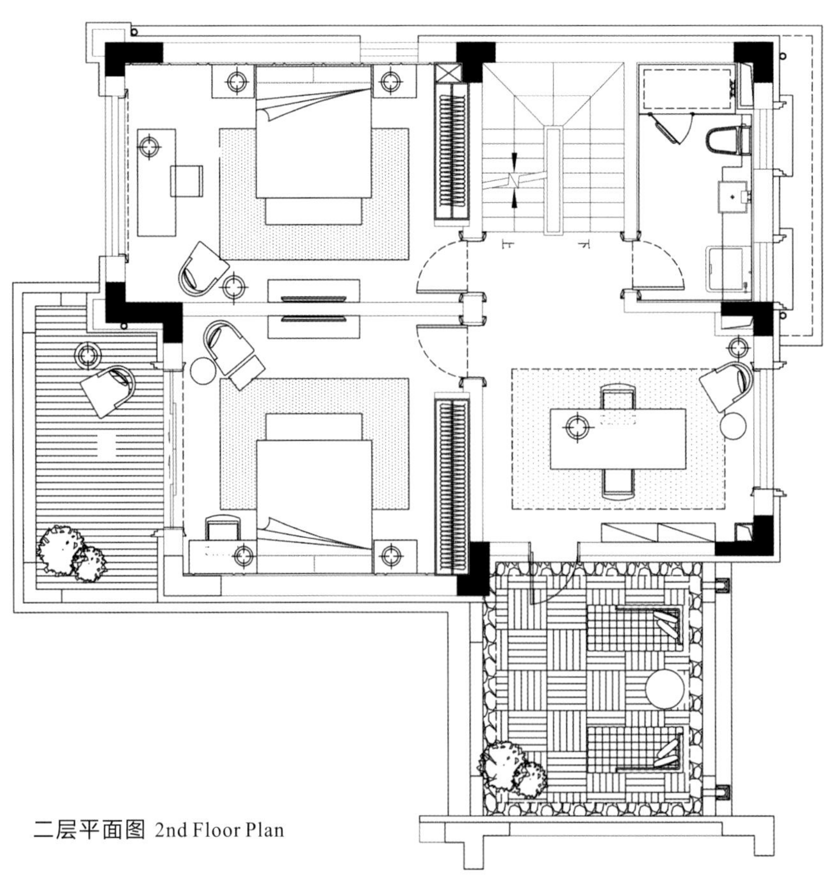

二层平面图 2nd Floor Plan

让人不容忽视的是，空间中精致的饰品有着画龙点睛的作用，水晶吊灯的点缀为空间锦上添花，这样细腻的设计给业主带来了宾至如归的感受。

The delicate decorations play a noticeable role in the whole project. For example, crystal chandelier makes people feel like at home.

## 色彩搭配

## COLOUR COLLOCATION

该案作为一套法式新古典的大户型，整体色调以浅色系为主，空间简约、唯美，让人能感受到美好、恬淡的生活。设计师在家具、窗帘、饰品、床品等细节元素上都采用了浅色系。从家具的色彩选配中流露出业主不喜欢被束缚的个性和高雅的生活品位。如客厅整体用色淡雅，以浅色系为主；其他区域一致选用米色的流苏布艺沙发，其外形简洁、线条流畅，既舒适又有气质；白色和蓝色的布艺虽然清新素雅，却点亮了整体空间的色彩；简易壁炉及背景墙搭配金色挂镜，以最自然而真实的语言“诉说”着业主对高品质生活的追求。

As a set of big house in French neoclassical style, the case is mainly in light color to create a beautiful and peaceful atmosphere. All the furniture, curtains, decorations, bedding and other details are in light color, show the personality and the taste of the owners. The whole living room looks elegant in light colors. Creamy white sofa is simple and comfortable, white and blue cloth is fresh and elegant, and fireplace and gold hanging mirror are showing the owners' pursuit of high-quality life.

SOUL A COASTER
R.KELLY

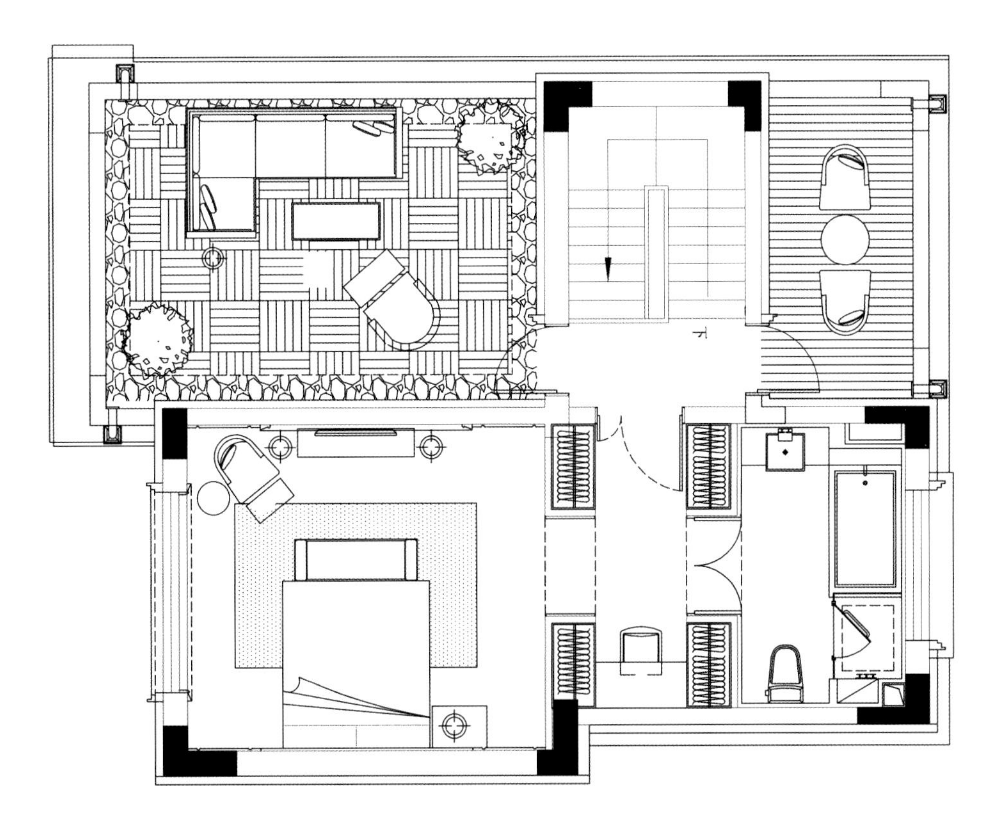

三层平面图 3rd Floor Plan

# LUXURIOUS NEOCLASSICAL CREATING HAPPINESS SPACE

## 奢华新古典 缔造幸福空间

根据客户群的定位，设计师把该案定位为奢华尊贵、沉稳大气的新古典风格。新古典风格注重对细节的要求，同时反映业主优雅、高贵的生活目标。

According to customer positioning, the case is positioned as a luxurious, noble and peaceful neoclassical design. Neoclassical designs pay attention to detail requirements, while reflecting the elegant, noble life goals of the owners.

## 家具配置

## FURNITURE CONFIGURATION

这个家到处充满着设计师的智慧：整体上采用不锈钢、木质、布艺、壁纸、玻璃、大理石，在展现尊贵感的同时还带有现代时尚感和舒适感。客厅里，通透的空间中摆放着柔软的布艺沙发和舒适的床榻，以及相当人性化的靠椅和大气上档次的水晶吊灯；卧室中看似随意搭配的装饰画、精致的吊灯和配饰，让别致和惊喜随处可见。空间中家具组合形成的流畅的造型，就如业主一样，时而稳重，时而优雅，但骨子里仍然流淌着绅士般优雅的血液。在这样的空间里生活，日子被“分解”成一个个充满设计灵感的时尚元素。

This home is full of designer's wisdom, and carefully planned comfort atmosphere can be seen everywhere: the entire space using a combination of stainless steel, wood, fabric, wallpaper, glass, marble, showing an incomparable superiority of the political and business elite, but also with a sense of modern fashion and comfort. In living room, there are soft fabric sofas and comfortable couch, very user-friendly chairs and luxurious crystal chandeliers. Unique and surprising design can be seen everywhere in bedroom such as random mix of decorative paintings, delicate chandeliers and accessories, etc. The smooth shape formed by the combination of furniture, just like the owner's personality, sometimes sober, sometimes elegant, but still a gentleman inside. In this space, the days of life are divided into some full of design inspiration fashion elements.

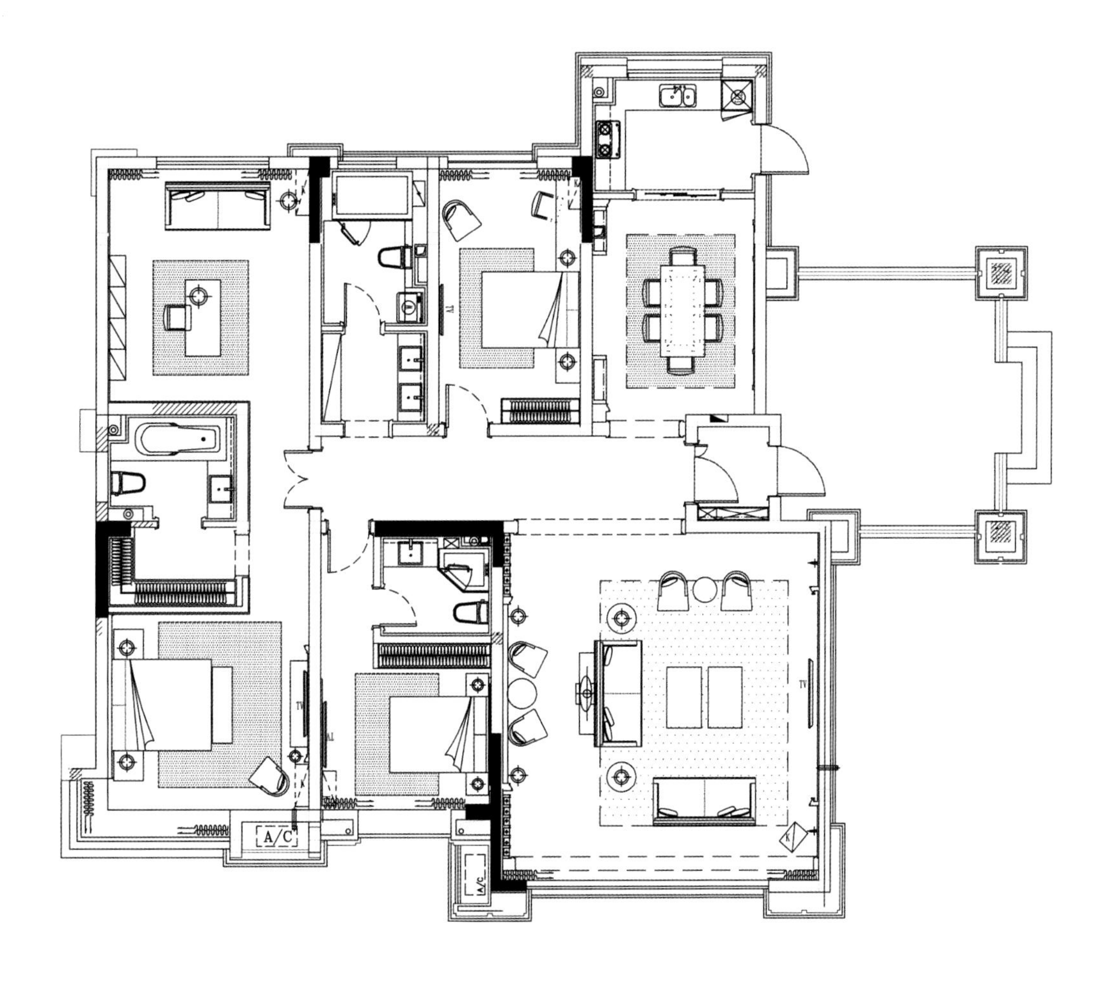
A/C

JOURNEES
PRISONNIER
Arsenal

THE ARSENAL
Emirates
Arsenal
JOURNEES
PRISONNIER
25 DECEMBRE
AU 1er JANVIER

## 色彩搭配

## COLOUR COLLOCATION

新古典对色调的要求极高，或是深咖色中适当地加以绿叶，或是金色中加以红叶作为点缀，一切搭配和调色都要突出现代奢华感。空间中色调丰富，木饰面以深咖色调为主色，搭配大理石材质、钢琴烤漆的质感及水晶质感的灯饰，红色雕花地毯在空间中成为点睛之笔，营造出贵气、典雅的空间。金属材质摆件和蓝色的床被色彩点缀让空间增加高贵气质和优雅感，精美、繁杂的画框和装饰画突显了主人公的尊贵。

Neoclassical design has high requirements on color tones. No matter dark coffee color properly mix and match some green leaves, or in golden tones using red leaves as embellishment, all color should highlight the modern luxurious feelings. In this colorful space, wood veneer with dark coffee color, collocate with marble material, plus piano paint crystal lighting, and outstanding red carved carpet, creating a noble and elegant space. Especially metal ornaments and brilliant blue color bed to add some noble and elegant atmosphere to space, exquisite and complex frame together with a significant decorative painting highlights the nobleness of the owner.

一个被时间装饰得有些古典的空间，散发着浓郁的生活气息。 This is classical space decorated by time, full of the passion for life.

# Elegant Neoclassical Full Of Noblest Atmosphere

## 高雅新古典 尽显贵族风情

本案摒弃过于复杂的机理与装饰，运用女性特有的视角观点及素材上的大胆搭配，将空间化繁为简地表现出来。大方的线条突出空间里的新古典意象，自然而绝对，呈现出业主的生活态度及品位追求。

This project abandons excessively complicated mechanism and decoration by using specific viewpoint of women and bold collocation in materials to make hard things simple. Decent lines highlight new classics images in the space, natural and absolute, exclusively presenting life attitude and taste of habitants.

## 家具配置

## FURNITURE CONFIGURATION

在家具的选择和搭配上，古典韵味最浓的要数客厅和卧室了。木质的墙壁和雕花壁纸，以虚拟画框的形式框出了一个个贴近自然的世界。质地舒适的沙发和素净的茶几，勾画出新古典的优雅与唯美，给人以开放、宽容、优雅的非凡气度，丝毫不显局促。卧室为平和而富有内涵的气韵及华美细腻的设计风格，拥有对称的新古典留白床头柜与其他家具，被自然光照射的饰品盒艺术品，更显温馨柔媚，突显出业主高雅的气质和贵族的身份。

The most classical places are Living room and bedroom. Woody wall with carving wallpaper frames natural worlds one after another. Comfortable sofas and purer tea table draw the outline of elegance and aestheticism of new classic, bringing out open, tolerant and elegant tolerance without cramped feeling. Walking into the bedroom, gentle and connotative artistic conception, gorgeous and exquisite design style, symmetrical new classical night table with white stripes and other furniture, jewel boxes and art works irradiated by light source, all of these make the space even warmer and sweeter at the same time tracing out elegant and noble identity of the owner.

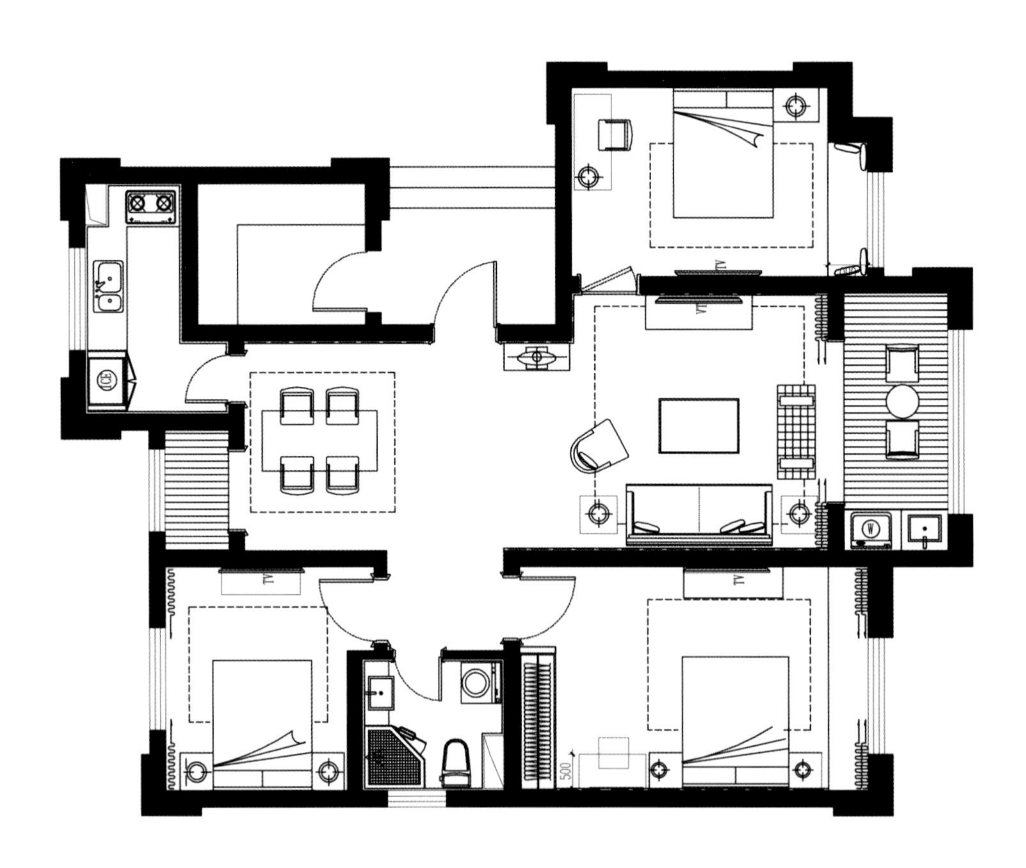
TV
TV
TV
TV
500

## 色彩搭配

## COLOUR COLLOCATION

为了使空间看起来明亮、大方、温馨，在色调搭配上设计师特别选择了白色、金色、黄色、粉紫为主色调。中性色与深色混合，打破了传统古典的忧郁和沉闷。如男孩房大红色的吊灯呈菱形，与床裙的色调相协调；黑色加黄色的台灯，流了一桌的“蛋黄”，尤其夺人眼球；金色壁纸提升了空间的质感，其他金色和白色的饰品与家具，使空间的色彩更加协调。

In order to let the space looks bright, decent and warm, designers particularly choose white, golden, yellow and purple as main tinge in color scheme to break melancholy and depression of traditional classic with a mixture of neutral and deep colors. For example, the bright red droplight in rhombus in boy's room coincides with the tinge of bed skirt; table lamp in black and yellow humorously likes “yolk” flowing on the table, catching people's attention; wallpaper with golden color modulation promotes spatial texture, with other golden or white decorations and furniture, balancing spatial saturation.

# LOVE FOR PASTORAL STYLE FEEL THE IDYLLIC SPACE

## 情动田园 诗意空间

本案没有华丽的装饰，亦没有过多复杂的造型，但设计师注重空间与人的和谐，在设计上将细节、人性化、舒适度在空间中恰当地融合，定制出具有充满幸福味道的田园空间。

Without luxuriant decorations and unique modeling, the designers focus on the harmony of space, integrating details, humane and comfort into one space, to customize the happiness pastoral space.

## 家具配置

## FURNITURE CONFIGURATION

家具主要采用与大自然接近的陈设和用材。客厅的真皮沙发、材质清新的电视柜和书桌、雅致的茶几等线条简洁明快；镶嵌在金色画框里的手绘画，时尚有趣；在布艺的选择上采用棉、麻布料，不用刻意去维护；这都与田园风格不饰雕琢的追求相契合，令空间散发出清新芬芳的田园气息。

Mainly use furniture made of natural materials. In living room, the leather sofa, fresh materials TV cabinet and desk, and elegant coffee table are all using clean and simple lines; the painting embedded in gold frame, fashion and full of fun; On the choice of fabric, using cotton and hemp cloth, those materials do not need to maintain, and fit with pastoral style which doesn't pursuit of complex decorations, and exude a fresh pastoral atmosphere.

## 色彩搭配

## COLOUR COLLOCATION

色彩最能直接表现空间品质并渲染环境氛围，绿色、米色、粉色、米白色，或深或浅、或淡或雅，通过设计师巧妙的搭配，打造了一个清新又温馨的空间。淡雅的窗帘、白色轻柔纱幔、蓝白精致的花器、素色却生动的动物摆件等装饰在保证了色彩统一的同时，增加了空间环境的丰富性。再加上清新的绿色植物巧妙地点缀在家里，让一切看起来生机勃勃。在这样充满诗意和快乐的空间里生活，幸福感会一直充满心间。

Color is helpful for creating environmental atmosphere. Designers use various color, green, beige, pink, white, dark or light, pastel or elegant, to create a fresh and intimate space. With elegant curtains, white soft Shaman, blue delicate flower pot, plain but vivid animal ornaments and other decorations maintaining a uniform color, and coupled with fresh green plants cleverly decorated at home, the house looks vivid and colorful. Living in this space, happiness will always be in your heart.

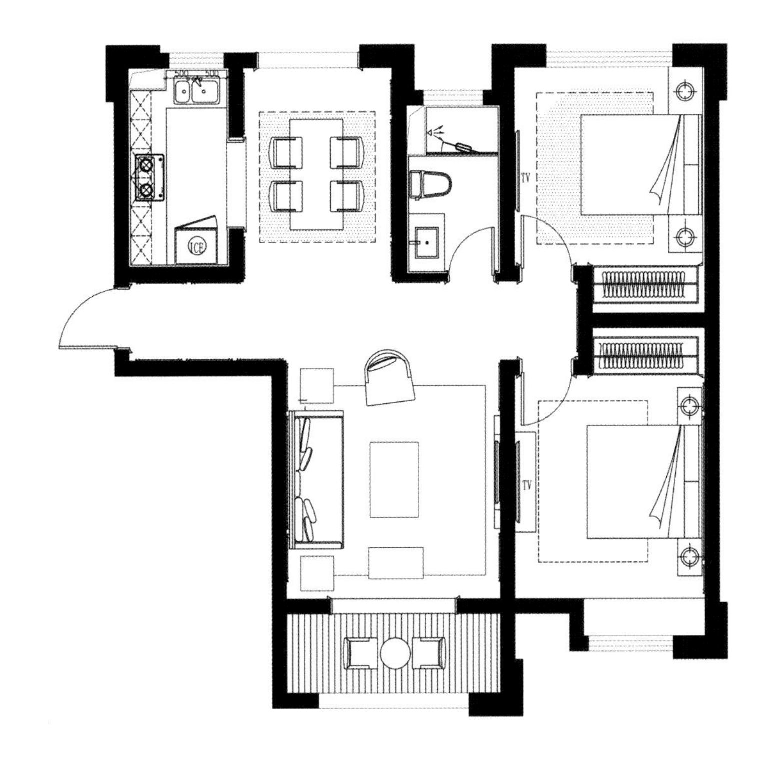


田园风格不是“风格”，而是一种取向，一种方式，一种对生活态度和生活理念的诠释。

Pastoral style is not only a "style", but also a life style, a attitude towards life and an interpretation of philosophy of life.

# MODERN WARM HOME SHOW FASHION AND PERSONALITY

## 现代温润居 尽显时尚与个性

无论是硬装还是软装设计，现代风格总是和时尚分不开。但该案恰恰相反，没有使用多么高科技的产品和国际潮流的元素。设计师经过对空间严谨地考量之后，用简练的语言诠释了一个现代之家。

The modern style is always about fashion. But the case is not the same. It does not use high-tech products and international trend elements, after a rigorous consideration of space designers use concise language to interpret a modern home.

## 家具配置

**FURNITURE CONFIGURATION**

从家具的选择上可以看出设计师对现代风格独特的理解。客厅里家具的组合在通透的空间中犹似一幅画。电视柜墙面延伸出的休息区，以电视墙面为隔断，把有限的空间视觉最大化，创造了小空间大视野的奇迹。两盏台灯分别置于多人沙发两旁，呈现出对称美。设计师还特别选用了几盆绿植，置于窗旁、墙角及楼梯口，突显不对称的居家美感及视觉效果。其他区域如卧室、餐厅、书房的窗帘、布艺、装饰画等的家具摆设和陈设品也较为简洁，沙发以单人椅为主，其他饰品为具有玻璃质感的金属材质。线条简洁的沙发与各种家具混搭，将现代风格软装的精神体现得淋漓尽致。

From the choice of furniture we can see the designer's unique understanding of the modern style. The combination furniture in living room forming a wonderful picture in this transparent space, TV cabinet wall extending a seating area to maximize limited space creates an effect of broad vision in small space.

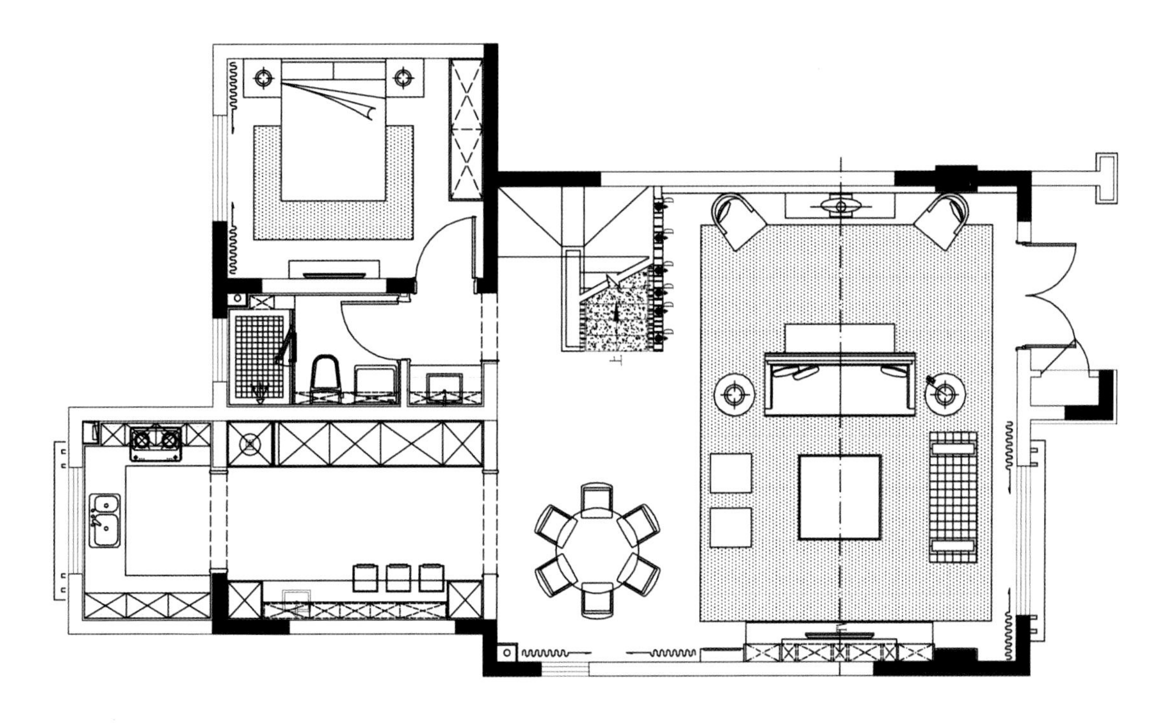

一层平面图 1st Floor Plan

## 色彩搭配

## COLOUR COLLOCATION

在色彩搭配上，无论是哪种风格的软装设计，讲究的都是业主个性品位的体现。该案的色彩搭配可用两个字形容——沉稳。现代风格下的沉稳别有一番特点。客厅的家具以含蓄、时尚的棕色带点亲近自然的褐色为空间主要色调，现代感极强的玻璃茶几在灯光照射下熠熠生辉；嫩绿的盆栽置于空间的点、线、面上，运用对比色打造视觉冲击。卧室棕色的床上有显眼的黄色抱枕。儿童房的人物装饰画色调丰富，跟床上的摆件色彩和谐一致，构成一幅独特的画面。

Soft decorations in every style should reflect the personalized taste of the owner in color. We can use one word to describe the color of this case “peace”. In the modern style, the “peace” will have a unique lingering charm. Take a look at the living room furniture, subtle and stylish brown a little close to the natural brown color set the tone for the space, with a strong modern flavor glass coffee table, gleaming in the light; the green potted plants are placed on the points, lines and surface, use colors to create a spatial vision. Like furniture selection of living room, the color of other living follows the living room, such as the brown bed matching with dazzling yellow pillows; the decorative paintings in the children's room are colorful and match the bed ornaments.

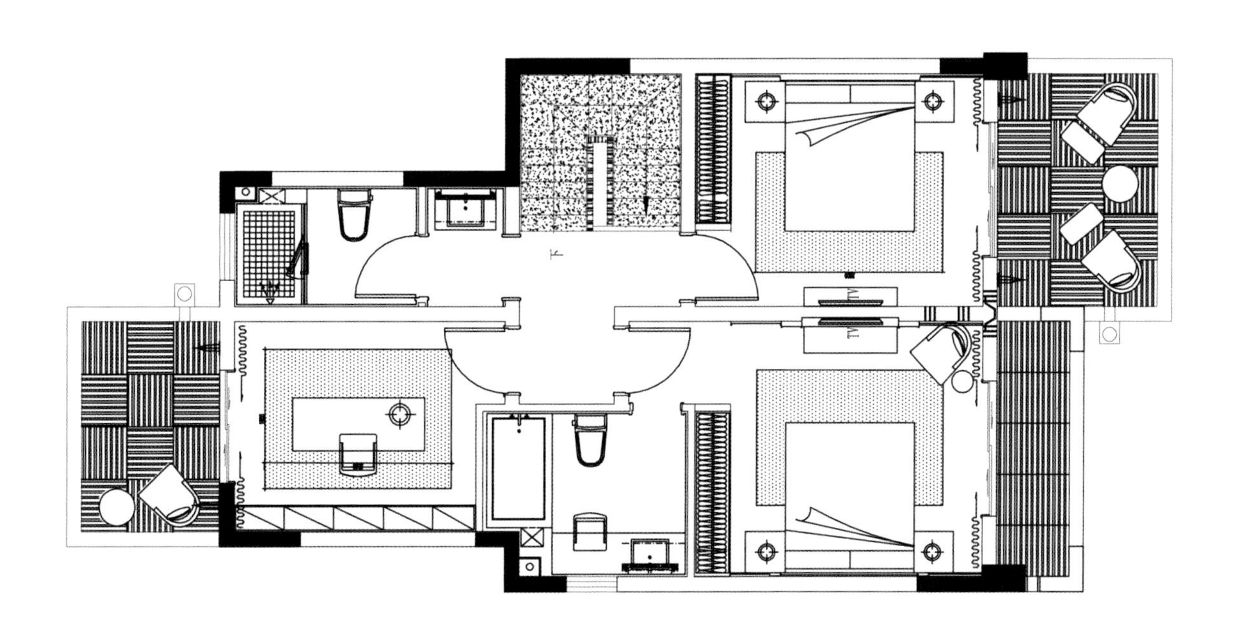

二层平面图 2nd Floor Plan

# MODEST DIGNITY
# ROYAL NEW LUXURY

## 低调的尊贵 皇家新奢华

“豪宅回归本质生活”是海珀系列一直秉持的宗旨。此次的海珀澜庭建筑风格为新皇家中式建筑风格，运用新中式艺术造型经典元素，继承中国传统建筑文化底蕴及设计精髓，打造空间层次变化丰富的立体空间。为了契合硬装的风格特点，软装设计师把空间定位为国际奢华风格，让皇室华贵的空间内涵表现得淋漓尽致。

" Returning to the essence of life " , this is the concept of Sea Amber series always persistent. This time the architectural style of Sea Amber Lanting adopts new royal Chinese architectural style, using the classic elements of the new Chinese art form, inheriting the traditional Chinese architectural culture and design essence, to create dimensional space with rich changes. In order to fit the hard mounted style features, soft decoration designers position the space as an international luxury style, letting this luxurious royal space to be wonderful.

## 家具配置

## FURNITURE CONFIGURATION

低调却不乏奢华，是具有一定生活品位的人追求的格调。该案在家具的配置上，设计师将各个空间的家具融合统一，以低调奢华为主要风格。首先是规整的客厅，中式的挂画搭配充满意境的朽木装饰，中式的格子地毯延绵开来，浪漫的靠椅和舒适的沙发组合，茶几上的动物摆件在明亮吊灯的照射下活灵活现。其次是卧室的装饰，从主卧、老人房到儿童房，无论哪个空间都呈现出低调的奢华：贵气的布艺装饰、高档的法式床品、充分烘托出空间归属感的中式床头柜、现代艺术感极强的挂画，以及为不同空间定制打造不乏生活情趣的饰品摆件……在这栋奢华的豪宅里，既可以看到融合了法国、意大利等经典欧式的格调，又能够发现少许中式的元素。

Low-key but without shortage of luxury, is what a person with a certain quality of life pursuit of. In this furniture collocation case, designers integrate and reunify the furniture of each space, with the concept of modest luxury. The first is the regular Chinese-style paintings and dead wood decoration in living room, together with Chinese plaid carpet, romantic combination chairs and comfortable sofas, animal ornaments on the coffee table in the bright chandelier come alive; then to the bedroom decoration, from the main bedroom, elder room, boy's room and to the girl's room, no matter which space, extravagance decorative fabric, valuable and tasteful French bedding, Chinese bedside cabinet space with a sense of belonging, modern art paintings, and custom interesting decorations for different spaces ... in this luxurious house, you can see a blend of French, Italian and other classic European style, but also be able to find a little taste of Chinese.

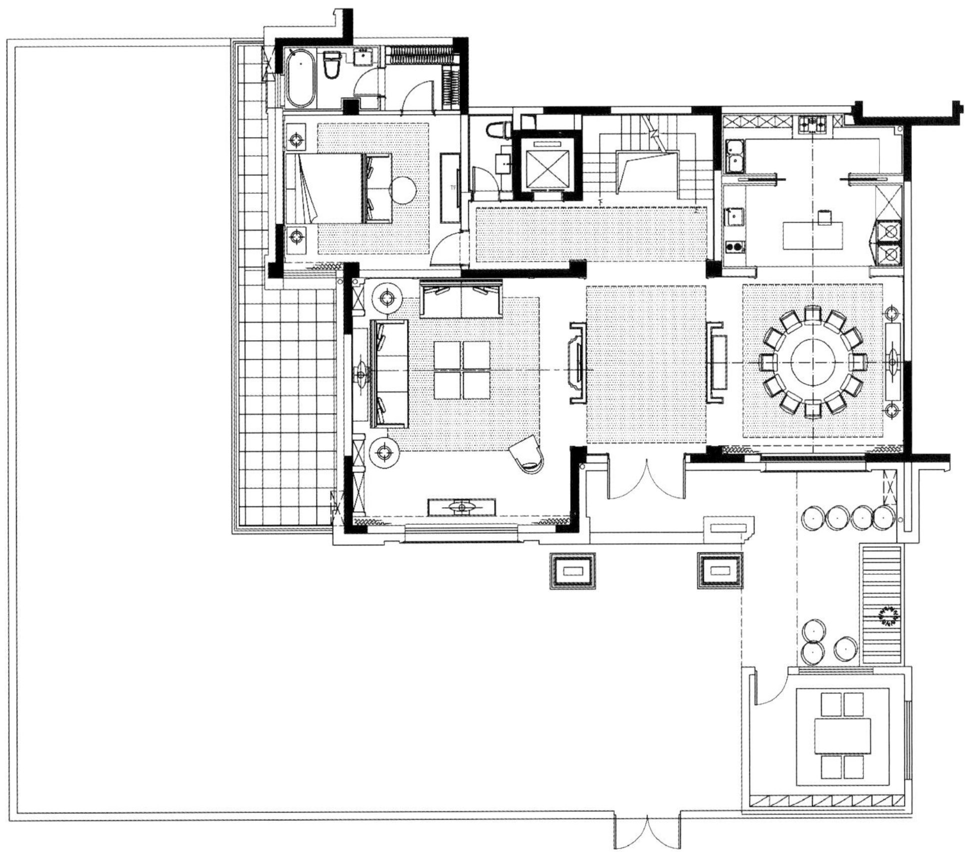

一层平面图 1st Floor Plan

## 色彩搭配

## COLOUR COLLOCATION

深棕、麻灰、米白构成了这个居所的主要色调，辅以一些嫩黄、紫红、深蓝等欢快的亮色，使空间色调层次丰富多样。黄色是一种强烈而醒目的颜色，通常为公共空间的重要标志色调。在客厅中，拿捏得当的嫩黄搭配深棕的装饰柜及红色的动物摆件，其温馨的色调与灯光交相辉映，让人一走进空间就觉得眼前一亮。相对于客厅的色彩搭配，卧室的色调跳跃度很高。客卧呼应整体空间的色调，以浅黄色为主色调，搭配深蓝色，如深蓝色的抱枕和座椅搭配米色的床被及嫩黄的壁纸；主卧中米白色和深棕色搭配的雕花床品，扑面而来的是一股贵族气息。主卧中最吸引眼球的是深红的床头柜，让人感受到浓郁的中式韵味。其他空间如男孩房或书房中，大红、米白、深棕等承载着视觉特质的色彩的艺术品，每种颜色和材质都相互衬托，散发着不同的能量。

The main chromatography of this residence is consist of dark brown, heather grey and off-white, which is accompanied by some lively bright colors like bright yellow, purplish red and dark blue, to make tinge gradation of the space abundant and divers. Yellow is a kind of intense and striking color which is usually used for significant symbol in public spaces. In the living room, appropriately nuanced bright yellow combined with dark brown sideboards and red animal ornaments reflected by cozy tinge and lights, let the space jump into your view when you step in. Compared with color matching in the living room, bedrooms have a strong leaping in tinge. Bedrooms respond the overall spatial tinge, matched navy blue basing on pale yellow as dominant hue. For example, navy blue bolsters and chairs go with cream-colored quilts and bright yellow wallpapers; carving beddings with off-white and dark brown matching in master bedroom sprays boundless nobility directly in your face. In fact, the most attractive element in master bedroom is the dark red night table which emits full-bodied Chinese charm. Colors including bright red, off-white or dark brown bearing visual idiosyncrasy in other rooms as boy rooms or the study, collocated with carefully selected artworks, emanate different energy and feeling relieved with every color and texture.

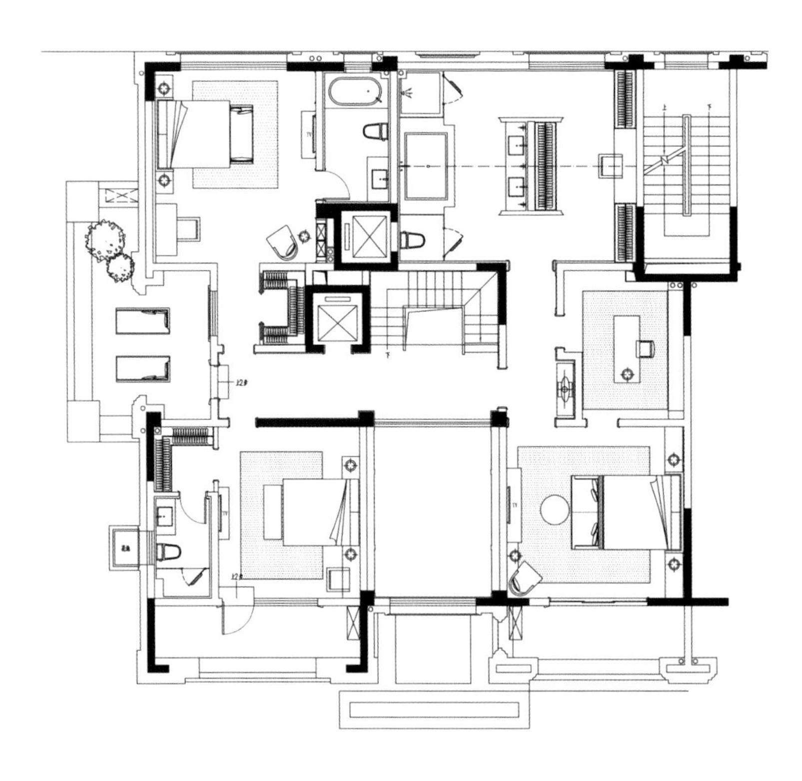

二层平面图 2nd Floor Plan

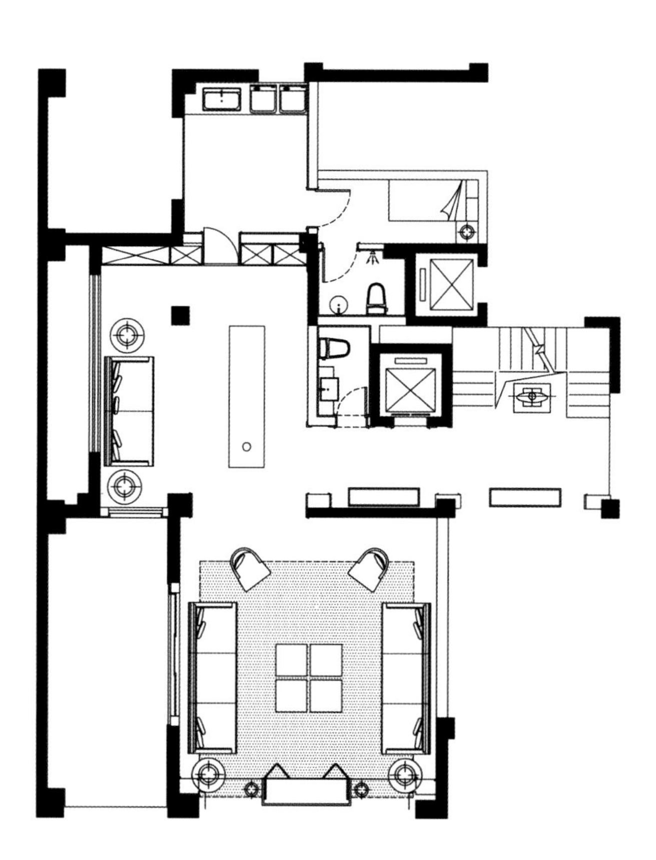

负二层平面图 -2nd Basement Plan

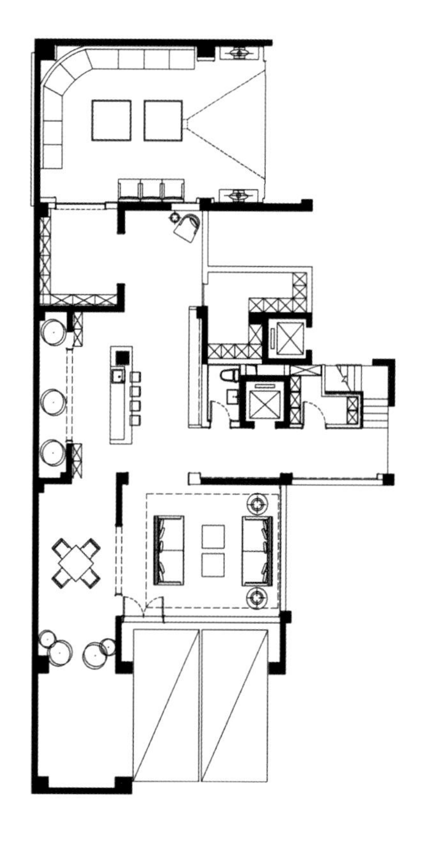

负一层平面图 -1st Basement Plan

# Charming Royal Purple Profound Dignity

## 魅力蓝紫 深邃的高贵

绿地昆明海珀澜庭代表了皇室贵族无上的尊荣、恒久与奢雅，在设计上突破了空间的局限，将东西方的经典美学融汇在一起。

Greenland Sea Amber Lanting represents the supreme honor, long-lasting and elegant luxury of royal nobles. And designed to break the limitations of time and space, mix the classic aesthetics of East and West together.

## 家具配置

## FURNITURE CONFIGURATION

该案被定义为大都会风格，家具均以直线略带弧线为主，并采用质量上乘的布艺、不锈钢等材质，营造出极为人性化的低调奢华、舒适的居所，体现出业主精致的生活理念。客厅的家具实用性较强，没有多余的装饰品，柔软、舒适的沙发就是客厅的主角。其他区域如休息室、活动室等的空间软装设计，无论是具有设计感的家具，还是精致的配饰，都被设计师融合在一起，体现了业主对高品质生活的追求。

另外，少量的金属和现代面料为空间增加了不少时尚、大气的感觉。或是镶嵌在墙上的花鸟图，或是挂于墙上的抽象画，这些个性的装饰画给人带来一场视觉上的享受。同时，进口金属质地的饰品和水晶艺术品，也让空间多了一份艺术内涵。

The case is defined as a modern luxury style, so the furniture is mainly designed to be straight lines with slightly curved, using high-quality fabric, stainless steel and other materials, creating a very modest luxury and comfortable accommodation, providing a delicate living concept for customers. The furniture in Living room has a strong practical performance, without extra decorations, soft and comfortable sofa is the leading role here. Other areas such as rest room and activity room, either the stylish furniture or delicate decorations are all mixed in one space by designers, to express the owners expectation of quality life.

In addition, a small quantity of metal together with modern fabrics adds a lot of fashion sense to the space. The picture of bird and flower embedded in the wall, or abstract paintings hanging on the wall, all these personalized decorative paintings bring a visual enjoyment for people. Meanwhile, the import metal decorations and crystal art works also add some art elements to this space.

三层平面图 3rd Floor Plan

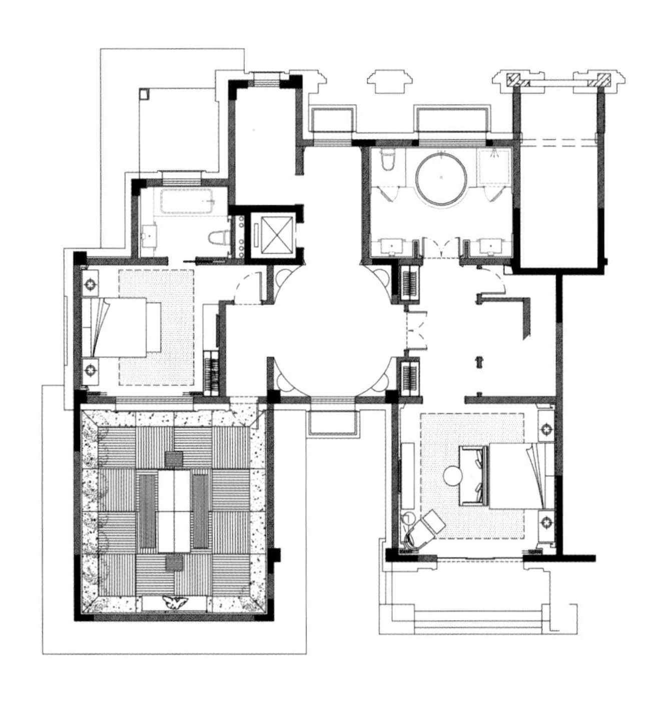

四层平面图 4th Floor Plan

## 色彩搭配

## COLOUR COLLOCATION

为了使空间带给业主如酒店式的精致和舒适，设计师把空间主色调定为蓝紫色。蓝紫色调既显浪漫与梦幻，又有华贵高雅的意蕴。客厅和活动室中的蓝色沙发显得格外抢眼，地毯的大幅雕花、深咖啡色的墙面散发的光泽，仿佛“诉说”着一场浪漫的“邂逅”。而主卧和阁楼书房以深咖色的木饰面为主色，辅以黄色或大红色；少量的金属工艺，搭配蓝紫色系的高级布料和摆件；让空间充满品质感和层次感，营造出一个色调丰富又突显奢华的氛围。

In order to make an exquisite and comfortable hotel style space for owners, designers decide to use royal purple as the main color of this space. Royal purple is full of romance and fantasy, with the charm of luxury and elegance. Blue sofas in living room and activity room are particularly eye-catching, sharply carved carpet and dark coffee shining walls all in the telling of a romantic encounter. The main bedroom and loft study room using dark coffee wood color, and complemented by yellow or red, collocate with a small amount of metal crafts, plus royal purple high quality fabrics and ornaments, to make this space filled with a sense of quality and layering, creating an atmosphere not only has abundant tones but also highlights luxury.

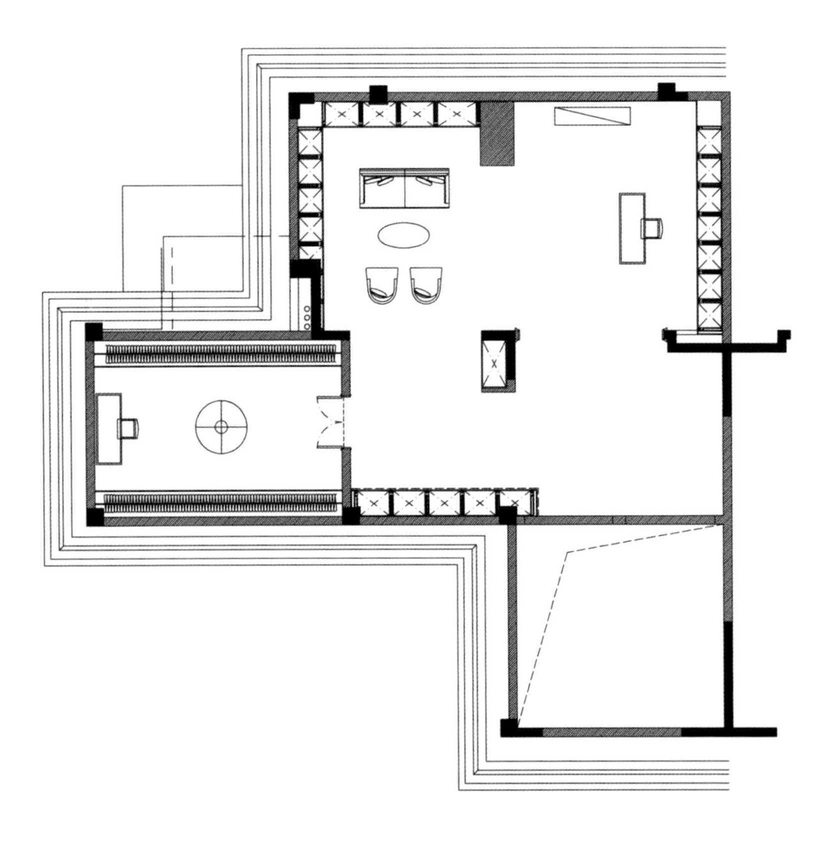

阁楼平面图 Loft Plan

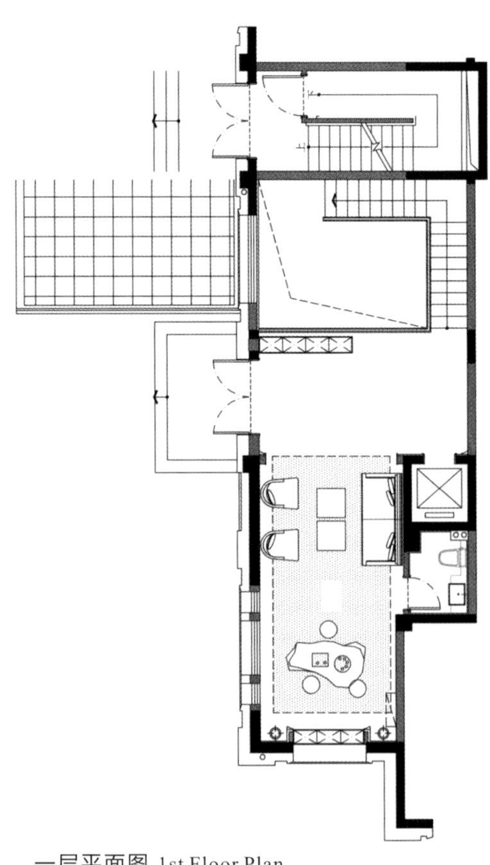

一层平面图 1st Floor Plan

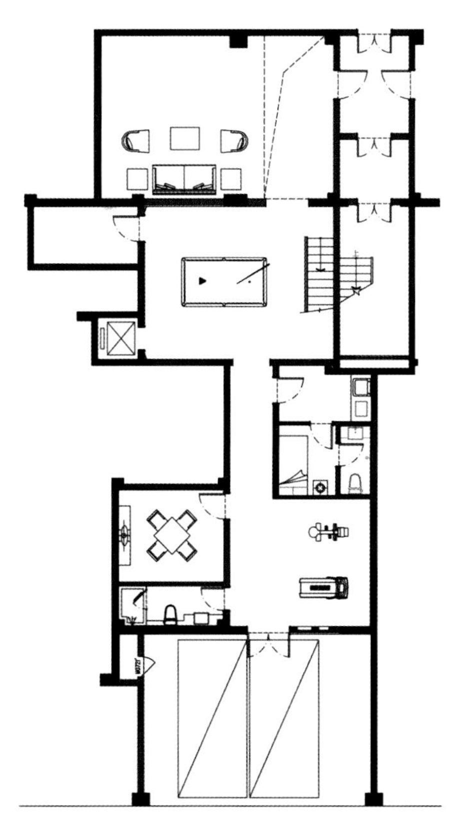

负二层平面图 -2nd Basement Plan

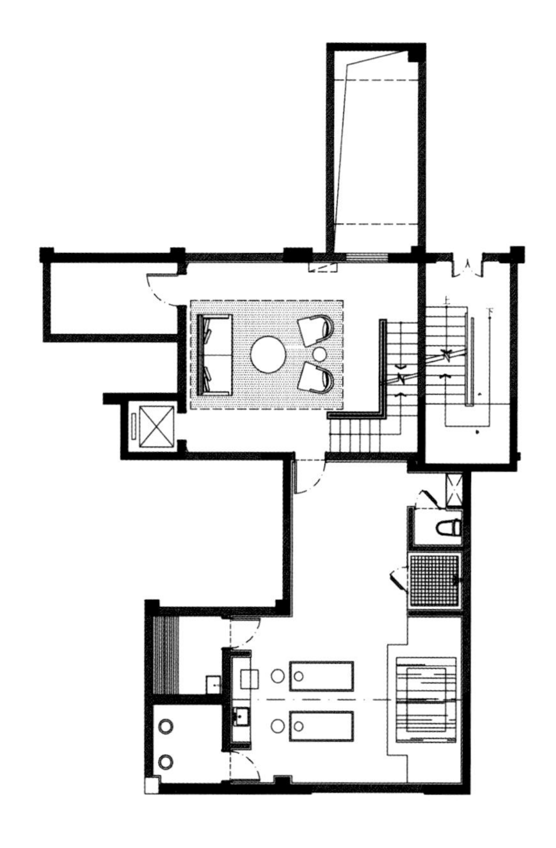

负一层平面图 -1st Basement Plan

# DARK COFFEE TONE CREATING MODERN LUXURY

## 深咖格调 打造现代轻奢华

销售中心考虑到公司与客户、与人群之间的“连接和互动”的重要性，以及空间中的材质选择、色彩搭配、造型设计，特意结合铆钉装饰，突显精致、细腻的美，再加上丰富的布艺和现代的工艺品，打造了一个引领时尚的现代空间。

Considering about the importance of "connectivity and interaction" between the sales center, customers, and crowd. And in space design, also consider the aspects of materials, color matching, and shape-designing, combined with rivets decoration, with exquisite sense of volume to highlight the delicate beauty, plus rich fabrics and modern arts and crafts, to create a leading fashion modern space.

## 家具配置

**FURNITURE CONFIGURATION**

洽谈区分为两部分，可为客户提供不同的舒适度，在让空间更加有层次的同时让客户与客户之间更有隐私感，表现了雅致、奢华又富有灵性的空间格调。在满足销售需求的同时，为更进一步增强客户的体验感受，设计师在许多细节上进行了深入的推敲，在家具方面利用不锈钢和大理石相结合，让整个空间简单而不失稳重。不同面料之间的相互碰撞让整个空间更为之高档。这样整体的搭配让销售中心更加时尚、简约又不失奢华。

The discussion area divides space into two parts, providing customer different kinds of comfort, and make this space more structured and more privacy, showing an elegant luxury and rich spiritual space. To meet sales demand, while for further enhancing the customer experience, designers make a deep scrutiny in many details. The use of stainless steel and marble furniture make the entire space simple yet stable; and the collision between the fabrics let the whole space full of quality. The overall combinations make the sales center more fashion, simple and luxury as well.

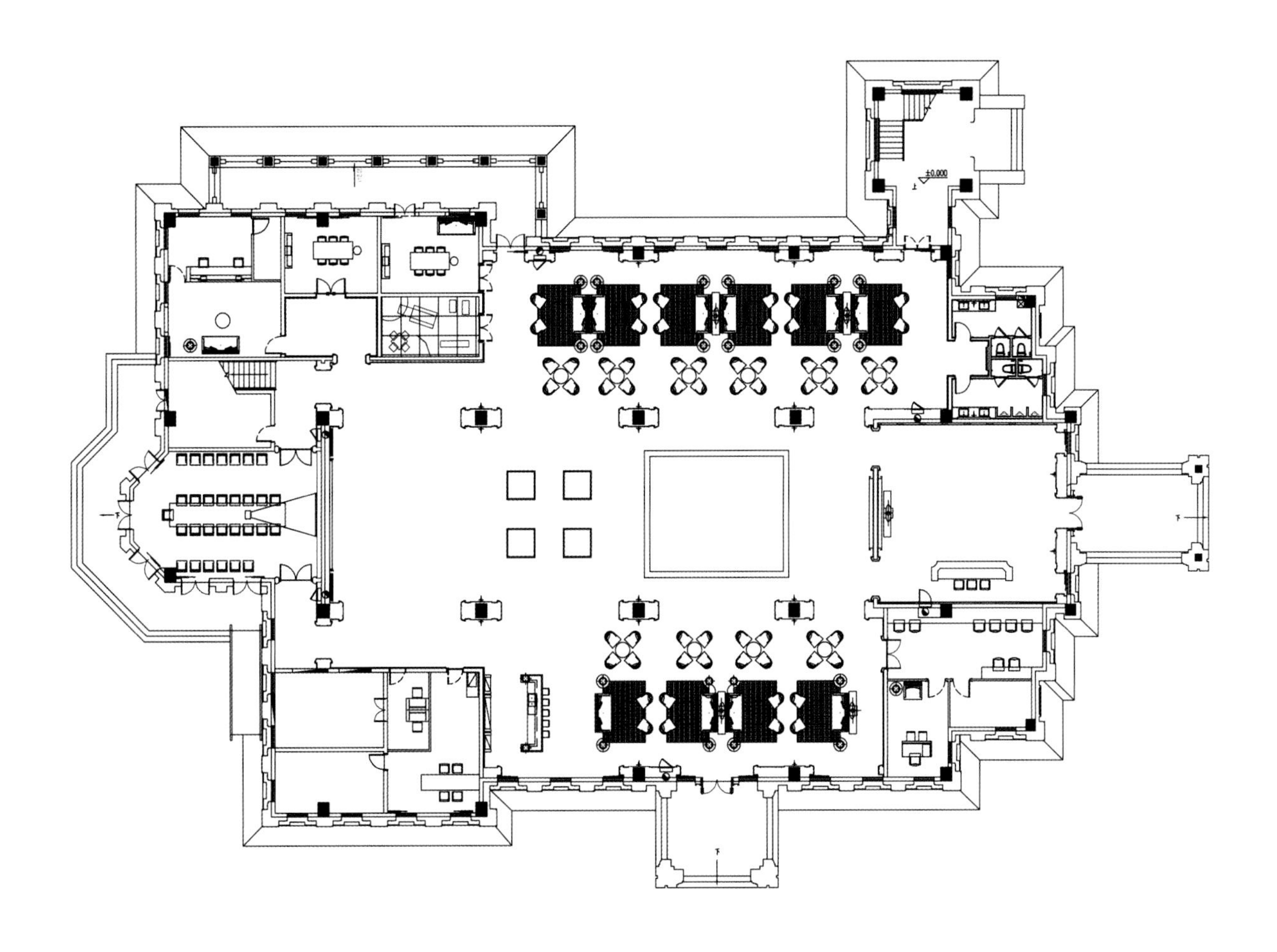
±0.000

## 色彩搭配

### COLOUR COLLOCATION

米白色、咖啡色、金色是该案的主要色调，洽谈区米白色的双人沙发和咖啡色的单人靠椅，搭配黑色的茶几，这种经典的组合，打造了一种特征明显的现代风格。最吸引眼球的是儿童区，销售中心安排了一个供儿童玩乐的空间：丰富的色调将儿童带入了一个七彩的世界，深咖啡色花纹的布艺、各种儿童玩具和摆设让小孩仿佛进入了一个大型游乐园，在这个色彩缤纷的空间里，一家几口的家庭可以有亲子之间的互动。这样的设计让客户更觉得贴心和细心。

Creamy-white, coffee color, gold are the main colors of this case. The creamy-white sofa and coffee color single chairs in signing area collocate with black coffee table, this classic color combination to create an overwhelming modern style. The children's area is the most eye-catching place. The rich colors give the child a colorful world, plus dark coffee pattern fabrics, all kinds of toys and decorations make children seem to have entered a large amusement park. In this colorful space families can interact with each other, feeling even more intimate and considerate.

# NOBLE AND FASHION SIMPLE EUROPEAN SPACE

## 时尚尊贵的简欧空间

本案融汇了中式与西式的风格，使得西方建筑的欧式简约与东方的审美方式完美结合。空间气氛充分体现了“时尚尊贵”的主题，软装设计除了考虑客户的时尚审美，更关注亲切与舒适的体验感。

The fusion of international style makes the simple European architectural perfectly combine with oriental aesthetic, fully expressing the “noble fashion" theme of this space. Soft decorations consider not only the fashion aesthetic of customers, but also more concern about intimate and comfortable experience.

## 家具配置

## FURNITURE CONFIGURATION

由于销售中心不同于样板间或者其他酒店空间，因此，它对软装设计的要求也不同。作为一个公共空间，在家具上主要以舒适的沙发座椅、丰富的布艺、现代的工艺品等来打造一个引领时尚潮流的现代空间。

洽谈区是客户看完房子累了可以休息和谈话的地方，主要针对一些有明显购房意向且不想被打扰的客户，所以设计师选择了椅背较高的单位沙发家具，这样就能有更好的私密性，并采用了舒适度高的面料。而活动区的家具与其他区域有别，这里的家具的摆设有一定的距离，保证一定的私密性。洽谈区与儿童区非常接近，所以与很适合家庭式的客户进行洽谈。

Sales centers differ from show house or hotel space, so the soft decoration requirements are also different. As a public space, we mainly uses comfortable sofa and chairs, rich fabrics and other modern crafts to create a modern space which leading fashion trends.

Discussion area is a place where customers can have a rest and talk with each other when they are tired and mainly for customers who are obviously intent to buy a house and do not want to be disturbed. So designers choose a higher back chairs and sofa to keep privacy, and also use high quality fabrics to maintain comfort. The furniture in activity area is different from other areas. Furniture placing here has a certain distance in order to keep privacy, and it's close to the children's area, to be suitable for family customer to negotiate.

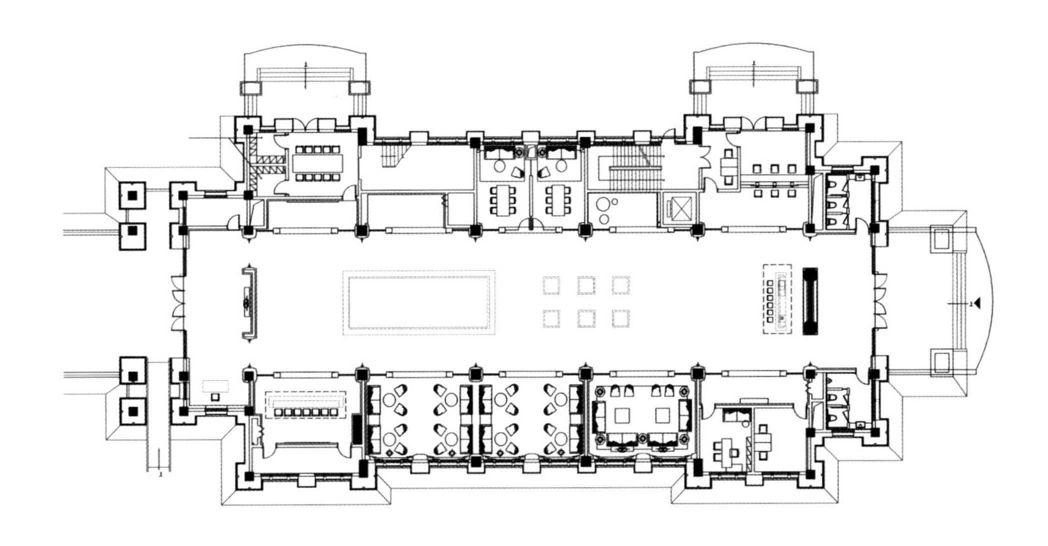

## 色彩搭配

## COLOUR COLLOCATION

空间色彩与线条充分表现了现代简欧风格，红色、黑色、白色就是这个空间的主色调，设计师利用这三种经典色彩的搭配来营造销售中心典雅的氛围。空间中悠然垂下的窗帘帘头、黑色的布幔与沙发上红色的抱枕、造型独特的灯饰、质地高档舒适的淡色布艺，以及刻意烘托氛围的灯光效果等搭配元素，从颜色和质感上让销售中心呈现出高水准的美，让客户在此尽情享受，整个身心沉浸在放松、舒适的氛围中。

Space colors and lines show the simple modern Europe form. Red, black and white are the main colors of this space. Designers use these three classic colors to create an elegant sales center. Leisurely hanging valance with a black curtain, collocating with red pillows on sofa, unique styling lighting, high quality comfortable fabrics, as well as lighting effects to heighten the atmosphere, all those elements from the colors to textures make this sales center present a high standard of beauty, making customers enjoy this place and putting their heart into this relaxed and comfortable atmosphere.

值得一提的是VIP签约室。这里是客户与商家进行签约的地方。为了拥有更好的私密性，设计师以窗帘作为屏障把VIP签约室单独隔出来。VIP签约室的整体装修相对洽谈区更简洁、大方、正式，客户在这里既可以签订文件，又可以在沙发上休息和进行交谈。

Especially the VIP signing room. Here is the signing area for customers and merchants, in order to have better privacy, designers use curtains as a barrier to separate it out, the overall decoration is simple, free and formal than discussion area. People can either sign documents or have a rest and talk with each other on the sofa here.

# LUXURY EUROPEAN STYLE SIMPLE BUT ELEGANT

## 奢华欧式 简练脱俗

欧式空间总是容易让人想起华贵，身在其中会有种提升身份的感觉。因此，强烈的欧式风格与现代的审美情趣也总是被设计师作为设计的出发点。该案通过对各个家具构件的组合处理及不同色彩的运用，大大地增加了欧式空间的包容性。

Space of European style are luxury and people feel noble in them. Therefore, intense European style and contemporary aesthetic taste are always regarded as the starting point of design. This project immensely increases the inclusivity of European space with its combination of soft decoration design and each furniture component in space, as well as application of different colors.

## 家具配置

## FURNITURE CONFIGURATION

作为一个销售中心，家具的选择总是有别于其他居住空间的设计，因此，对软装设计师的要求也更高。该案宽敞的空间为软装设计提供了很大的空间，专门为客户量身定制的舒适家具，如单人椅或双人沙发，满足了不同客户的要求。水晶吊灯透亮、大气，是空间的一大亮点。大面积的装饰画散发出强烈的艺术气息，抽象的画面带给客户无穷的想象。令人难忘的还有墙角的一些物件—— 波纹状的花器呈现出的动感，带来意想不到的效果。

As a sales center, furniture selection is different from other living spaces. Therefore, the request to soft decoration designers is higher. The capacious space of this project provides prodigious space to soft decoration design which includes customized and comfortable furniture for customers, single chairs or double sofas, to meet the requests of different customers. One of highlights of the space is the bright and decent crystal chandelier. A large area of decorative picture reek with strong feeling of art, and its abstract image brings guests with endless imagination. Some articles in the corner that bring unexpected effects with dynamic presented from undulatory floral organs.

## 色彩搭配

## COLOUR COLLOCATION

任何一种风格的软装设计，色调的搭配都具有至关重要的作用。色彩不但能调和空间的美感，还能提升空间的品质。该案以清新的米白色和稳重的棕色为主色调，米白色的双人沙发和靠椅两两组合，质量上乘的黑色茶几与其高度相同。棕色的单人靠椅很人性化，质感舒适突显华贵。这样的色彩搭配给人一种错乱美。而绿色植物在该案中被大量地运用，如地板上、茶几上、墙面上、挂画下。绿色植物在空间中无处不在，一股清新浪漫的氛围扑面而来。尤为抢眼的是入口处的拱形门上，由玫瑰、马蹄莲等多种花材组合成的那一圈白绿相间的花儿，以清新的景致欢迎着四方来客。

Color is very important in soft decoration design of any styles. Colors can decide how the space looks like and feels like. Fresh creamy white and steady brown are used as main tinges in the project. The color matching that double sofa in creamy white and arm-chair combine with each other and hold the line with superior quality tea table in black with humanized brown chair and luxury texture brings deranged beauty of link and split. Green plants are largely used in the project on floor, tea table, wall space or under pictures. The fresh and romantic atmosphere sprays into your face when you see green plants everywhere in the space, especially at the arch door of entrance where various flowers like roses and callalilies make up a white-green-interphase ring that welcomes visitors in an overlooking pose.

## 广州市汉意堂室内装饰有限公司
## Guangzhou Haniton Decoration Design Co., Ltd

自2006年成立以来，广州市汉意堂室内装饰有限公司以优质软装设计闻名软装业界，积极拓展业务至全国30多个城市，设计范围包括样板房、酒店、会所、别墅、商业空间软装设计、家具设计、装饰画设计、床品布艺设计、花艺设计，多年来备受推崇。汉意堂总部拥有将近3000$m^2$的设计、研发基地，汇集了几十位国内外设计精英，组成强大的设计与实施团队，拥有出色的设计开发能力及市场整合能力，为国内多家大型地产商及五星酒店缔造了一个个经典软装工程案例。公司总部位于中国广州，在美国洛杉矶及中国北京、西安、济南、苏州等地设有办事处，公司对团队的陈设设计素质要求极为严谨，所有设计由广州总部负责，分公司与旗下机构则肩负筹划、联络、协调等事务，确保为客户提供专业的、高水准的软装配套服务。

公司注重软装作品的生活哲学细节化与时尚价值艺术化的完美结合，从设计之初到工程实施完工，力图将品质生活提升到哲学高度，高屋建瓴，以艺术创作的思维去进行渲染与彰显，表达出强大的艺术张力，创造出震撼的视觉效果。

Since its establishment in 2006, Haniton Decoration Design (HDD for short), well-known for high quality of soft decoration design, has expanded its business from Guangzhou to over 30 other cities around China. The scope of its business includes the design of show houses, hotels, chambers, villas, business space arrangement, furniture, decorative paintings, floriculture, bedding & cloth art etc. The headquarters of HDD has a design and research center of about 3000 square meters and dozens of domestic and overseas design elite. Its strength in design & development and its sense of the market have created many classic projects for large scale property developers and five-star hotels. HDD situates its headquarters in Guangzhou and establishes branches in many other cities like Los Angeles, Beijing, Xi'an, Ji'nan and Suzhou. HDD is known for its meticulous requirements for the quality of its team and outcome. To ensure its high standard in interior design provision, the headquarters of HDD is in charge of all the design of its projects. Its branches and related organizations are responsible for the planning, communication and coordination of specific issues.

From the design of the project to final complement, HDD pays attention to both life philosophy and fashion, tries to add artistic value into common life, and creates amazing visual effects.

01 | 02 | 03 | 04

01. 公司标牌 02. 公司大门 03. 公司大门拐角 04. 口号

享受软

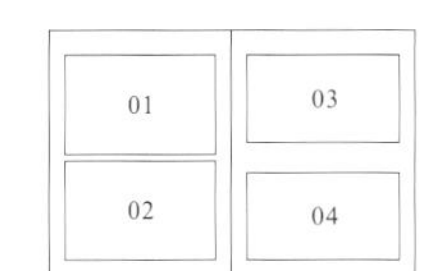


01. 公司大院　02. 树屋　03. 设计工作室　04. 绘画研究室

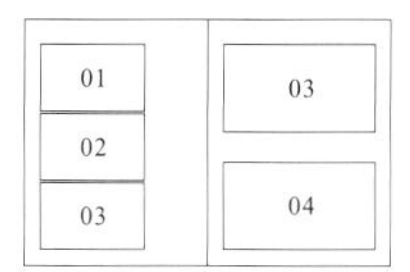


01/02/03. 总监办公室　04/05. 布艺研究室

01. 多功能厅　02. 过道　03. 洗手间

# 企业荣誉

# Enterprise Honor

**2009 南方都市报最佳软装奖**

**2010 亚太地区“筑巢奖”空间比赛银奖**

**2011 “筑巢奖”酒店空间比赛银奖**

**2012 年度国际空间设计大奖——“艾特奖”最佳陈设艺术设计入围奖**

**2013 第八届中外酒店白金奖最佳酒店软装设计机构**

**2013 中国家饰行业年度杰出软装设计品牌**

**2014 广东省建筑软装饰行业协会监事单位**

**2014 广东省陈设艺术协会理事单位**

2009 Southern Metropolis Daily, Best Soft Decoration Award

2010 Asia-Pacific Region "Nest Award" Space Design Competition, Silver Award

2011 "Nest Award" Hotel Space Design Competition, Silver Award

2012 International Space Design Award -- Idea-Tops Best Design Award of Art Display, Nomination Prize

2013 The 8th International Hotel Platinum Award, Best Hotel Soft Outfit Design Agency

2013 Chinese Furniture Industry Annual Outstanding Soft Decoration Brand

2014 Guangdong Construction and Decoration Materials Association Supervisor Unit

2014 Standing Member of Association of Art Crafts and Decoration Industry

原研哉
Kenya Hara
国际著名设计师
日本设计中心代表
武藏野美术大学教授
日本平面设计师协会副会长

**筑巢奖国际评委原研哉评语：**

“中国的追求”形象化后以这样的形态展现在了我们的眼前。“中国的现代主义”仿佛是在接受了各种所求后才得以自由地发展壮大。原本在装设物品方面就具有极高造诣的中国，再加上最新照明技术的支持，使空间环境愈发凝造出一种沉练与简洁。

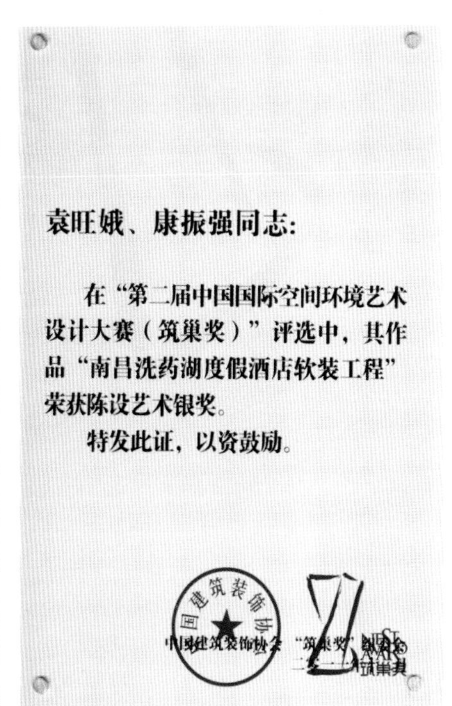

**袁旺娥、康振强同志：**

在“第二届中国国际空间环境艺术设计大赛（筑巢奖）”评选中，其作品“南昌洗药湖度假酒店软装工程”荣获陈设艺术银奖。

特发此证，以资鼓励。

中国建筑装饰协会　“筑巢奖”组委会

## 作者介绍
## Editor Profile

**袁旺娥**
**Wanda**

**个人简介：**

出生于海南三亚

毕业于中国美术学院视觉传达系

米兰理工大学室内设计硕士

海南省女画家协会会员

海南省三亚市摄影协会理事

广州市汉意堂室内装饰有限公司董事兼设计总监

Introduction:

Born in Sanya, Hainan province

Graduated from China Academy of Art

Interior Design Master's degree of Polytechnic University of Milan

Member of Hainan Female Painter Association

Director of Hainan Photography Association

President and design director of Haniton

**康振强**
**Joey**

**个人简介：**

出生于山东临沂

毕业于中国美术学院国画人物系

广东省建筑软装饰协会监事

广州市汉意堂室内装饰有限公司总经理兼艺术总监

广东陈设艺术协会、中装美艺、江西美术专修学院庐山特训营等多家

软装培训机构专家讲师

Introduction:

Born in Linyi, Shandong province

Graduated from department of Chinese figure painting of China Academy of Art

Guangdong Construction and Decoration Materials Association Supervisor Unit

The current general manager of Haiton

Expert lecturer of many soft decoration training organizations, such as Guangdong Association of Art Crafts and Decoration Industry, Chinese Soft Outfit Education and Lushan training camp of Jiangxi Academy of Fine Arts, etc

## 采访人介绍
## Interviewer Profile

**黄浩罡**
**Hocking**

**个人简介：**

毕业于广州美院

品牌形象设计师、著名设计传媒人

广州群英汇文化传播有限公司 / 创办人 / 董事长

国内首个互动设计管理课程ECC-DMBA / 创办人

广州国际设计周设计师俱乐部 / 总经理

2014世界室内设计大会（中国广州）组委会委员

金堂奖评审委员会委员

微信互动平台：ECC群英汇 / 出品人 / 总策划

Introduction:

Graduated from the Guangzhou Academy of Fine Arts

Brand image designer, famous design & media worker

Founder and chairman of Eliteshow Communication Co., Ltd

Founder of the first interactive design management course ECC-DMBA in China

General Manager of designer club of Guangzhou Design Week

Member of the Organizing Committee of 2014 World Interiors Meeting in Guangzhou

Member of the Jintang Prize Committee

Weixin interactive platform: Producer and general planner of ECC Eliteshow

**康振强、袁旺娥业界荣誉：**

2006　绘画作品被《生活速递》收录刊登

2009　荣获南方都市报“完美我家”空间设计大赛最佳软装奖

2010　以大连鲁能优山美地V3号别墅软装工程获“筑巢奖”陈设空间类银奖

2010　大连鲁能优山美地V3号别墅软装工程被《中国建筑装饰装修》总第88期收录刊登

2011　以南昌洗药湖五星度假酒店软装工程获“筑巢奖”陈设艺术银奖，
　　　日本无印良品设计总监原研哉亲自为其写了评语。

2012　南昌洗药湖五星度假酒店软装工程被《雅居生活》收录刊登

2012　以西安香树花城一期花园洋房B1户型软装工程获“艾特奖”最佳陈设艺术设计入围奖

2013　参与《软装设计教程》写作并出版

2013　参与《跟着大师学软装》写作并出版

2013　《软装设计手册》特约点评专家

Honors:

2006　Paintings were published in *Life Express* magazine

2009　Won the Southern Metropolis Daily " Perfect My Home " Space Design Competition Award for best soft decoration

2010　Luneng Yosemite V3 villa won " Nest Award " Space Design Competition, Silver Award

2010　Luneng Yosemite V3 villa was reported in *Interior Architecture of China* magazine

2011　5-star Green Lake Resort in Nanchang won " Nest Award " Art Display Silver Award, and the MUJI design director Kenyahara wrote comments for her in person.

2012　5-star Green Lake Resort in Nanchang reported in *Soft-Living* magazine

2012　Cree and Flower of the City first phrase B1 garden house in Xian won Idea-Tops award, the Nomination Prize of Best Design Award of Art Display

2013　Published the book of *Soft Furnishing Design Guide*

2013　Published the book of *Learn Soft Decoration Designs from Master*

2013　Special commentator of *Soft Decoration Manual*

策划：广州市汉意堂室内装饰有限公司

文案：潘富鸾

排版：胡莹

摄影：康振强、陈奇

Planning: Guangzhou Haniton Decoration Design Co., Ltd

Copy Writer: Pan Fuluan

Designer: Hu Yin

Photographer: Joey,Chen Qi

## 图书在版编目(CIP)数据

我为绿地做软装 / 袁旺娥，康振强编著 . －武汉 : 华中科技大学出版社, 2014.12
ISBN 978-7-5680-0447-3

Ⅰ. ①我… Ⅱ. ①袁… Ⅲ. ①室内装饰设计－作品集－中国－现代 Ⅳ. ①TU238

中国版本图书馆CIP数据核字(2014)第236346号

**我为绿地做软装** 袁旺娥 康振强 编著

出版发行：华中科技大学出版社（中国・武汉）
地　　址：武汉市武昌珞喻路1037号（邮编:430074）
出 版 人：阮海洪

责任编辑：刘锐桢　　责任监印：秦　英
责任校对：杨　睿　　装帧设计：张　靖

印　　刷：北京利丰雅高长城印刷有限公司
开　　本：889 mm × 1194 mm　1/16
印　　张：16
字　　数：128千字
版　　次：2014年12月第1版第1次印刷
定　　价：258.00元（USD 57.99）

投稿热线：(010)64155588-8815
本书若有印装质量问题，请向出版社营销中心调换
全国免费服务热线：400-6679-118 竭诚为您服务